Polar Hole Light in Europe's Clouds

The Asimov-Sagan Axis Shift and Mars Collapse Puzzles

Ruth Leedy Carr

PAGE PUBLISHING
Conneaut Lake, PA

First originally published by Page Publishing 2024

ISBN 979-8-88960-810-3 (pbk)
ISBN 979-8-88960-823-3 (digital)

Printed in the United States of America

Contents

Introduction

When Carl Sagan put "Earth axis shift" and "field reversal" and a cataclysmic author's name into "Stefan Alexeivich Baruda" in his novel *Contact*, was that just dumb luck? Was it just luck that Baruda delivered a fiery doomsday warning at a time when public fears of doomsday and even *an Earth axis shift* had been roused by a message from outer space? I had sent Sagan two of my books warning of an Earth axis shift as an outcome of an approaching geomagnetic field reversal. My full name then was Ruth Anne Leedy, which can be found in Baruda's name (as long as Leedy is respelled as Leedi). Just luck?

Try looking in fictional character names for the word axis, and see how easy it is to find just that one word. It's almost impossible. Sagan's knowledge of codes shines through in his novel, and so he was no stranger to anagram messages. Was it luck that his prisoner name, Tyrone Free, spells, "Eye R not free"? How about his book title that sounds like, "broke his brain"?

Isaac Asimov didn't just accidentally put "Planet Earth be hollow" in both Winthrop Carver Cabwell and Hortense Hepzibah Lowot from his story, "Perfectly Formal." Lowot is "frontally concave," as a planetary shell would be internally concave. And it was no accident that the names Merry, Tessa, and Ranay from Asimov's novel *Nemesis*, where an unknown star is heading for Earth, turned up in the letters adjacent to commas and apostrophes of his later story, "Frustration." He was signaling to us.

"Tar Heel, North Carolina," *may be an anagram*, according to the plot of Isaac Asimov's mystery, "Irrelevance!" When we notice that the first five letters can spell Earth, the name definitely starts to look like a puzzle. I see the following possible message: "On the hollo Earth theori, I cannot tell anione the trooth."

Our scientists have told us the difficult truths we rely on them to tell us. Now all we need to do is learn to read.

Chapter 1

Secret Sun Smoking Gun from Isaac Asimov; Mars Polar Light Photos—More Smoking Guns

A *secret star* is near us. Many photos confirm this. If only scientists would tell us so...

But Isaac Asimov has told you so. All you have to do is write down the letters before and after every comma and apostrophe in his story "Frustration" (34), and you will find three names from his novel *Nemesis* about a secret star threatening Earth. When he inserted "Merry," then later "Tessa" and "Ranay," into a random list of letters next to punctuation, he was counting on someone to be looking for a secret message like this—someone who had *already* found many other covert signs in his work. The three names clearly mean one thing:

"Read my book about the secret star." This message tells us that the novel *Nemesis* is describing a *real* situation, a *real* secret. This is the clearest sign of the cover-up, but Asimov and Carl Sagan have provided many other signs: hints, riddles, and SOS signs. They did this at great personal risk and at a high cost. We will be looking at the retaliation they suffered.

NASA took the safe way out when my congressman requested their comments about polar light on the nightside of Mars. They refused to comment on the analysis I had provided, or on the photographic evidence itself. Their excuse for not providing such basic information was pretty weak.

It is the retouching of NASA's Mars photos of 1969 that gives away the fact that there is secrecy about the bright discs of polar light that we see in these photos. The day-night border at the right edge of the polar "cap" has been erased in various ways on each photo (see figures 3 to 6). Someone *didn't want you to see* that the polar "cap" was on the nightside.

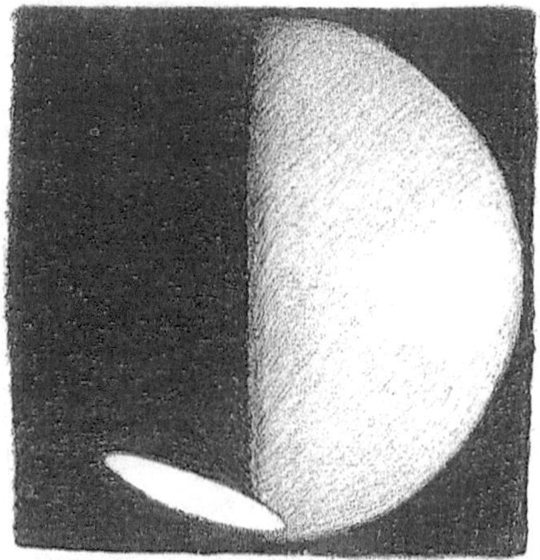

Figure 1. *Secret polar light source* shows up as being on the
nightside in this sketch of Mars based on photo C (figure 5).
That Mariner 7 photo has been extensively retouched to keep you
from noticing that the bright polar "cap" is on the nightside.

It is absurd in photo C for *two-thirds of the perimeter* of this Mars image to be illuminated, when exactly half of the perimeter should be visible when a planet is in half phase (or three-quarters phase or one-quarter phase).

Regardless of how the southern hemisphere on photo C is painted, the distance from the polar "cap" to the sun's bright reflection on the side of the planet should tell you that the "cap" is too far away from the sun's light to be on the dayside. If it isn't on the dayside, then it must have an independent source of light, especially considering how bright and extensive the polar light is.

Figure 2. This sketch shows how photo D must have looked prior to retouching (see figure 6). The planet had to be in less-than-half phase, since the picture was taken after the Mariner 7 probe had taken photos A through C, progressing from the sunny side to the dark side.

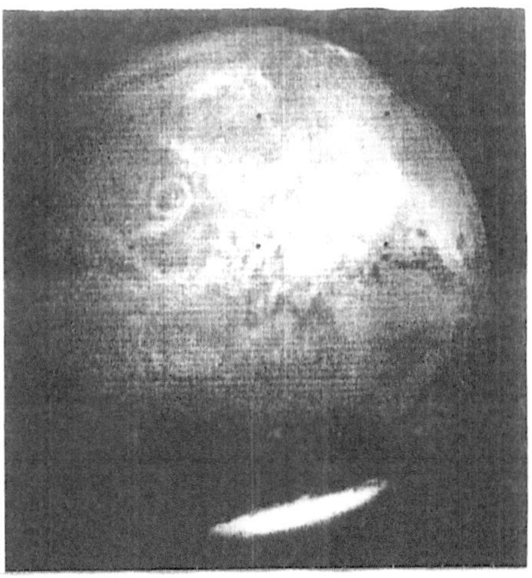

Figure 3. This Mariner 7 photograph of Mars, referred to here as photo A, is the first in a series of four pictures taken in sequence, all showing an extraordinarily bright south polar "cap" on the nightside. This picture is from *Space Science and Astronomy* (Page) and carries this caption: "Mars from Mariner 7 at 293,200 mi. distance, on

Aug. 4, 1969 at 10:38 UT with longitude 115 degrees centered. Nix Olympica (later renamed 'Olympus Mons') is the white circle ¾ in. toward upper left from center. Bright south-polar cap at bottom. (JPL-NASA photo.)" The planet is in about ¾ phase here, except that the southern hemisphere must have been artificially darkened.

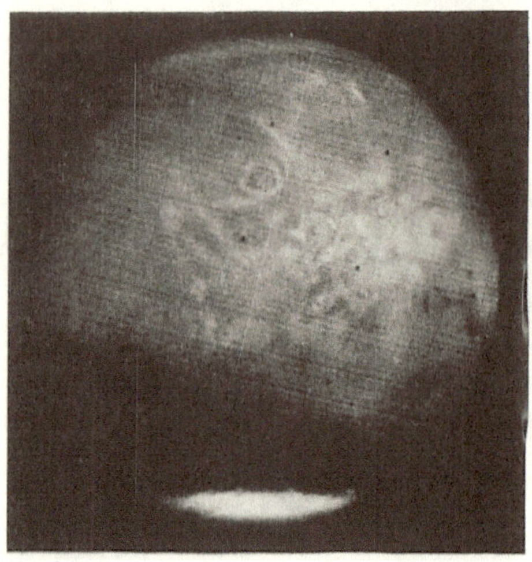

Figure 4. Referred to here as photo B, this Mariner 7 photograph of Mars, taken shortly after photo A, is from Patrick Moore's *Guide to Mars* (1978 edition). Both Olympus Mons and the sun's bright reflection are farther east in this picture, which is in about 2/3 phase (except for the artificial darkening of the southern hemisphere). A small dot from the dot matrix should be showing near the upper right edge of the polar "cap," but I foolishly whited it out some years ago so the polar "cap" would be uniformly bright as in photo A.

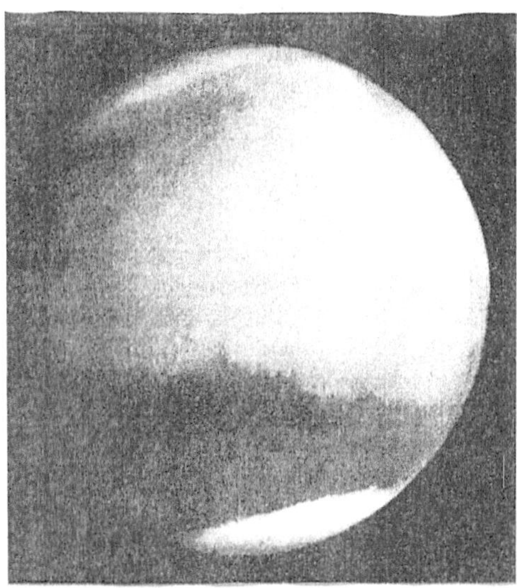

Figure 5. This Mariner 7 photograph of Mars, taken August 4, 1969, is from the *Encyclopedia Britannica*, 1974. This photo C, as we shall call it, shows a slight progression to the right of Olympus Mons and the sun's bright reflection, as compared to their positions in photo B. With illumination decreasing, this photo should appear darker overall than it does. It was artificially lightened, then the background sky was darkened, beginning at the glitch on the right side and extending below the polar "cap" and well beyond. The polar "cap" is thus artificially placed on the dayside.

Figure 6. This Mariner 7 photograph of Mars, which we will call photo D, appears on the title page of *Pictorial Guide to the Planets* (Jackson). As in photo C, there is artificial definition of the planetary outline extending under the polar "cap," that is, onto the nightside. The impropriety of this may explain why most of the "cap" has been cropped off. Though the brightest solar reflection point has disappeared from the right side, this photo has been made to appear as though brightness extends as far to the left as on photo A. Before retouching, the planet should clearly have been in less than half phase (see figure 2) because of the way Olympus Mons appears farther to the right than in photo C.

The tilting of the Image in photo C has been adjusted here in my figure 1 sketch so as to portray sunlight striking horizontally and creating a bright reflection halfway up. To portray sunlight shining down from above, as in photos A through C, is not realistic. When the image is adjusted so the reflection point is halfway up, the day-side must end at the top and bottom of the image, and the polar "cap" is clearly tilted away from the sun.

NASA has gone on record refusing a congressman's request to explain how the polar "cap" was illuminated on the nightside of photo B. The whole matter should be referred to scientific journals, the August 17, 2001, letter said. And yet the issue is obviously a hot potato. Because the rotation of Mars is faltering, as will be explained, the hidden sun illuminating the open south pole poses a grave danger to Earth. Consequently, it is up to *government* to show some leadership and explain the situation to the public.

The retouchers of photos A, B, and D portrayed darkness *all over* the lower southern hemisphere (except for the polar "cap") while in Photo C we see the appearance of dim *illumination* extending all across the lower southern hemisphere. Neither of these presentations is accurate. The sketches tell the hidden truth. The south polar lighting, since Mars has no auroras, is simply unexplained. Scientists have provided us with as much information as they were permitted to provide. We are lucky to have the pictures at all.

Shrugging off this issue as a curiosity would be a mistake. This photo analysis is the key to understanding several types of cataclysms that could be close on the horizon.

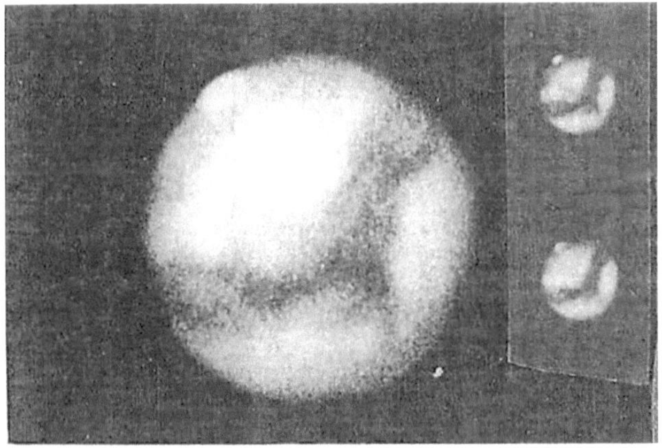

Figure 7. Light protrudes through polar openings, reflecting on dust or clouds, in these Mars photographs from 1902 and 1909. The larger one on the left is from George Gamow's book, *A Planet Called Earth*. The smaller ones on the right are from Marshall B. Gardner's book, *A Journey to the Earth's Interior*.

Marshall B. Gardner argued back in 1920 that the protruding blobs of polar light in those early photos of Mars were indicative of a central sun's light shining through polar openings. Even if there were solid ice caps on Mars, which have so far not been detected (only wet rocks), it would be hard to explain why we should expect such ice caps to produce protruding masses of light. George Gamow included one of these early Mars photos in his book, *A Planet Called Earth*. The caption calls the protruding blob of light an ice cap, though there was no proof of this. The text makes no mention of the picture, which speaks for itself. Gamow was very distinguished as one of the originators of the big bang theory.

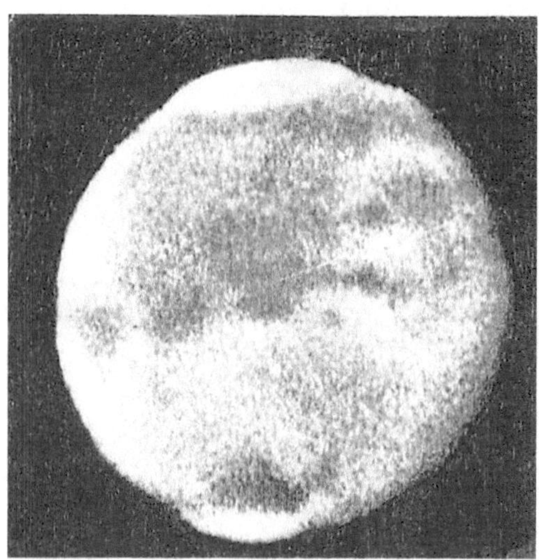

Figure 8. This extraordinary Mars photograph is from *Life Beyond the Earth* by Samuel Moffat and Elie A. Schneour. The caption suggests that future space missions will help us understand Mars better.

Two protruding blobs of polar light, one larger than the other, can be found on a Mars photo in *Life Beyond the Earth*, by Samuel Moffat and Elie A. Shneour (figure 8). If these were polar ice caps, we would expect that the existence of one fully illuminated polar cap would imply that the opposite cap should be in winter darkness totally and therefore not showing at all.

An apparent disc of polar light appears on the nightside of Earth in *Space Science and Astronomy* (Page), figure 133. The caption speaks of polar-aurora "caps" being visible on the nightside. The aurora, however, is a ring of light and not a solid cap, as can be seen here in figure 9.

Figure 9. The south polar aurora australis is shown to be
a ring of light in this photograph, published February 16,
1985, by *The Morning Call*, Allentown, Pennsylvania.

Although scientists have not openly defended the hollow planet theory, it is not fair to say they have not provided evidence to support it, as we have seen and will continue to see here.

Chapter 2

The Fate of Northern Europe, Japan, Hawaii, the Planet Mars, and the US West Coast: A Preview of the Riddles

Both Isaac Asimov and Carl Sagan teach us how to decipher their riddles, beginning with the easy and obvious and progressing from there. Before we retrace that learning process, and before we look at the scientific facts behind the cataclysmic predictions that you will be seeing, I want you to have a preview of the most dramatic messages these experts have provided.

"A polar hole shall end upper Europe soon as you and Ed Cayce prophesy, R. A. Leedy, perhaps four decades." The word "in" is missing from this Asimov riddle, probably because including the letter I would have made it impossible to decipher. The names on which this is based will be given later.

"You need to understand that a disaster's to destroy just the northern part of Europe." This, again, is from Asimov.

"A field reversal shall cause Earth's axis to be shifted (as Ruth Anne Leedy writes) in five, seven, nine decades." This riddle message from Sagan can end without including the portion in parentheses, but I believe he may have wanted to include that. There is a connection between Earth axis shifting and the character name from which this message was drawn.

"Upper Europe is to be a polar site. All Europe will be less warm. Most people must get out," (Asimov).

"Ruth L., you must help Europe's people soon to learn of my puzzles." When Asimov produced this riddle, I had sent him five of my books warning of an axis shift that would destroy northern Europe. The story from which this riddle was taken has a title that speaks of cold weather. The riddle mentioned just prior to this one comes from a story that involves *catching* a cold.

"Axis migration is near." This Asimov riddle comes from the name of a character who rides on a scooter. Each wheel of a scooter spins on an axis that migrates as the scooter moves along. It is nearly unheard of for Asimov's characters to ride on scooters.

"Land near the hole at Earth's axis is really going to shake in sixty to eighty years," (Asimov).

"A large hole near axis shall soon meander as axis meanders," (Asimov).

"The Earth shall be shaken as poles are reversed and polar holes recarved," (Asimov).

"In many reversals the axis remains the same," (Asimov). This tells us that if the new north magnetic pole moves to the same site as the old south magnetic pole and vice versa, the solar wind, in its path to those new magnetic poles, will not define a new axis.

"If the holes to the hollow of hell shifted, deaths would follow," (Asimov).

"Japan people are all endangered," (Asimov).

"France and San Francisco no are safe," (Asimov).

"Los Angeles no is safe," (Asimov).

"I want to warn Hawaii, Japan. That won't happen." Asimov explains in many riddles that he can't say what he would like to say because his brain has been tampered with.

"You say Earth is hollow, Ruth A. Leedy. You are right." This message can be found in the name of an Asimov character who has several things in common with me, as we will see.

"R. A. Leedy, you can learn my riddle game. You shall be my messenger." This is one of various riddles where Asimov appears to ask me to share his messages with the world and where he expresses a fear that this task might put me in danger. For years before I sent him my books, he was appealing in his riddles for help in sharing his cataclysmic warnings, as I hope to show.

"Mars shall end, and its inner star shall threaten the Earth," (Sagan).

"I, Asimov, say Mars is doomed and may end soon." The letters for "Asimov" are *grouped together*, but scrambled up, in the middle of the character name from which this message is drawn. That should be a *red flag* to us, indicating that the name contains a message.

"Trying to tell you Mars rotation is stalling is not going to be an easy job," (Asimov). I did figure this out.

"Mars be hollow." Asimov gave us this in Hober Mallows.

"Mars is doomed. I am scared," (Asimov).

"Mars end nears," (Asimov).

"Mars is not yet limp," (Asimov). The word totally could be added.

"Mars is about to go limp, to vibrate, crumple, collapse," (Asimov).

"Three planets are hollow," (Asimov).

"I am not able to say that Earth and Mars are hollow. My brain be bent and torn," (Asimov).

"My brain is disabled," (Asimov).

"My brain is not able to say, 'Mars is hollow.'" This riddle is from Carl Sagan, whose book title, *Broca's Brain*, sounds like, "broke his brain."

"Broca died, brain broken," (Sagan).

"You can see Broca's brain. It is broken and abraded," (Sagan). The brain of scientist Paul Broca is preserved in a jar, as Sagan relates in his book *Broca's Brain*.

"To rule the Earth, you need to learn to bend and tear the human brain," (Asimov).

"Our rulers are able to cause quakes, catastrophes. Our brains are not free, can be bent or broken," (Asimov).

"Sasquatch can cause quakes. You can be shaken," (Asimov).

"I are not free," (Sagan).

"I be not free. Brain be torn and abraded," (Sagan).

"In a puzzle, I can tell all." This is from Asimov's made-up word zapulniclate, beginning in "zapul," which easily converts to "a puzzl." Asimov did *another* made-up term containing "zapul." With these two terms, he declared himself a master puzzler, or riddler.

"Meet the real Riddler. I am he," (Asimov).

"I am labelled the Riddler," (Asimov).

"Hell has an alien civilization. It is not at all safe to tell of it," (Asimov).

"Why are you risking your life, R. A. Leedy?" (Asimov).

"Ruth Anne, you are fearless. Surely you realize Earth's king can break your brain," (Asimov).

"I didn't want you assassinated, Leedy, and so I didn't devise an easy test," (Asimov).

"I didn't intend to endanger you, Ruth A. Leedy, through any riddle that I did," (Asimov).

"I, Isaac Asimov, realize some will laugh at my revelations on Mars."

"Laugh if you want to, yet I know the hollow Earth theory is no joke," (Asimov).

"Do not let the Mars star land on Earth." A large orange ball lands badly in this Asimov story.

"If the Mars star falls on Earth, the Earth shall die," (Asimov).

"Mars internal star's light is visible to me, Isaac Asimov."

"Inner stars' rays are seen by scientists," (Asimov).

"Try to steer the Mars star, or it could destroy the Earth," (Asimov).

"Mars shell sags. Central star can chase and destroy the Earth," (Asimov).

"We are afraid to tell all to world. Ruler will retaliate," (Asimov).

"Ruler scares everyone in science, so all are living a lie," (Asimov).

"I worry over events that are to shake and even destroy the Earth seventy years down the road." This Asimov riddle contains the word seventy.

"I say the Earth has a hollow shell that orbits a little star at its center," (Asimov).

"We have a hollow world," (Asimov).

"Planet Earth be hollow." The letters for this sentence can be found in two romantically linked character names in one of Asimov's final stories, but not in any of his other character names that I know of. The bride-to-be is called frontally concave, reminding us of the concave inner surface of a hollow planet.

"Mars is doomed, and its inner star is on the verge of destroying the Earth." This is from Asimov's novel *Nemesis* in which a star is headed for Earth.

"Brain is disabled. I need your help. Decipher and share riddles." This is a 1955 message from Asimov.

"End-of-world riddles need reading. No one is willing," (Asimov).

"Each riddle I do is an SOS sign. I seek aid." This is from an Asimov character name, where the last name begins with "Sos." This message can be found in at least one other riddle.

"We cower below a ruler able to control a brain," (Asimov).

"Ruth Anne Leedy, you hold hollow Earth in your hand." This message can be found in at least two Asimov riddles.

"Bernard and Gardner be truthful. I lied," (Asimov). Bernard and Gardner are the names of two hollow-Earth theorists.

When the riddles are presented in more detail later, you will see that the spelling is not always perfect. Asimov himself used a respelling of Earth in the Dr. Urth mysteries. Another possible respelling of Earth would be "Irth," which is justified by words like mirth and birth. Asimov's character with the middle name Quackenbrane has the nickname Quackbrain. This tells us that "brain" can be respelled as "brane." In "The Fourth Homonym" Asimov points out that "to" can be respelled as "tu." Of course, a riddle is more believable if all the words are spelled correctly.

Solving riddles is not an exact science, but that is not necessarily a flaw in this form of communication. If the messages were totally clear, Asimov and Sagan could not have gotten away with putting so many riddles into the public record.

Chapter 3

"When It Rattled Across a Hollow World"—Isaac Asimov; Why the Earth Cannot Possibly Be Hollow

There is a quick way of proving the Earth cannot possibly be hollow, and this argument lurks in many people's minds. So before we look at any further hollow-planet evidence, we should check out the premises of this powerful, yet somewhat lazy, argument:

1. For the Earth to be hollow in our time, there would have to be a powerful cover-up.
2. A powerful cover-up would not permit a serious challenge to the solid planet model.
3. A hollow-Earth theorist who has produced books and newsletters since 1981 would be a serious challenge to the cover-up.
4. If the Earth were hollow, a theorist like Ruth A. Leedy Carr could not possibly exist.
5. Therefore, the survival of Ruth A. Leedy Carr proves that the Earth could not possibly be hollow.

In other words, my survival since 1981 may seem to some people to be proof that there is no cover-up over the hollow-planet question; therefore, the Earth is not hollow. Before we make that grand assumption, let us consider some reasons why I may have been allowed to survive. Let us say the high king Oz is approached by his chief henchman, who thinks it is time to silence Ruth A. Leedy. The year is 1986.

H: Just say the word, Chief, and I will see to it that this hollow-earther Ruth Liddy, Leedy, whatever, is rubbed out and forgotten.

Oz: I appreciate your concern, Hench, but you need to look at the big picture. There have been rumors and whispers of a hollow Earth for centuries, and none of it has made any difference. On the other hand, if a pattern develops where every hollow-Earth theorist drops out of sight, the issue might be taken more seriously.

H: You could be right. You usually are. Marshall Gardner didn't do anything after his 1920 book, his second effort, I believe.

Oz: A sad case. Something about a descent into madness. Only a fool would promote the hollow Earth theory in the face of united opposition from society at large.

H: It was too bad about Raymond Bernard. He did one spectacular book and not much else. He may have disappeared through a polar hole and settled in the interior, or so it was said.

Oz: Well, after his book came out in hardback and paperback, it was important for his writing career to come to an abrupt end regardless of what people might say. But Ruth Leedy isn't on that level at all. Now that her publisher, Gray Barker, has met his untimely end, she has no one to amplify her feeble efforts on her manual typewriter. So many typos. It's hard to believe a journalism graduate would be so lousy at proofreading.

H: The whole project is obviously overwhelming to her. And yet my sources report she has scored some points with her homemade books.

Oz: Homemade and lovingly glued together with—yuk—gallons and gallons of rubber cement. She's going to kill herself with that garbage, and we won't have to lift a finger. She just did a huge gluing project on her third book. Yuk!

H: But really, Chief, your puzzle masters are telling me that Isaac Asimov and Carl Sagan have done riddles in response to her work—covert endorsements, as it were.

Oz: I know all about the riddles. Do you know what the odds are that she'll ever discover the riddles? And even if she did, you can't prove the solutions are correct. And even if one or two riddle messages could be proven, is anybody going to listen to that kind of evidence?

H: So you are content to let this loose cannon keep firing…

Oz: This lady is pathetic, scarcely worth killing. She has no science degree. People who know her well consider her to be a little crazy. She had a nervous breakdown at the age of sixteen and never really pulled out of the fantasy world she constructed at that time. She is tortured by her TMJ jaw problem, which is only aggravated by these dumb book projects. She has burned out as a church organist and piano teacher and is going to leave her precarious home base in a trailer in Delaware to live with her parents in Florida. She is spiraling into oblivion without any need for a push from us.

H: I believe you feel sorry for this peculiar dame, Chief. Or maybe you enjoy listening to her piano-playing.

Oz: I like to govern with a light touch, Hench. You know that. Also, I enjoy a little harmless entertainment now and then. Instead of killing every potential gladiator, I enjoy watching them fence with the lions for a while if it is clear what the outcome will be. After all, this could be a bit of a test of my iron grip on the scientific community. In no way are they able to endorse Leedy's work openly, no matter what she says, now or ever. So where is the threat?

H: As you say, Chief, not much of a threat.

Another quick way to dispose of the hollow planet theory is even simpler than the first method. It goes like this:

1. Our scientists are free to speak and write as they please.
2. If the Earth were hollow (or facing an axis shift), this would not be kept secret by all scientists.

3. Therefore, the Earth cannot possibly be hollow (or be facing an axis shift).

This reasoning betrays a complete ignorance of how mind control can be exercised. It is possible today for *specific* thoughts to be weakened or suppressed without affecting other thoughts. These techniques have been developed to aid those who suffer from traumatic memories. Most memories and thoughts remain intact and can be exercised freely, but the targeted thoughts and memories can be suppressed. This is only one potential technique for preventing scientists from speaking out on the hollow planet theory. An array of other threats and assaults could be used.

Another popular method for disposing of the hollow planet theory is based purely on self-protection:

1. I am not obligated to get involved in any issue that could cause my career or personal relationships to suffer.
2. If I support the hollow planet theory in any way, I am likely to suffer some harm.
3. Therefore, I am entitled to go on thinking the Earth is solid without looking at any evidence as long as that is the prevailing view.

This argument is unassailable, since the right to free speech includes the right to think as we please. However, this type of thinking is synonymous with unquestioning belief and should not be confused with the type of thinking that involves the examination of evidence.

I sympathize with those who would like to dismiss the hollow planet theory with little or no serious thought or investigation. I had to force myself to think about it at first. It was painful, distasteful. I pushed myself to consider what Raymond Bernard was saying as an exercise in journalistic objectivity. If I had not been trained as a journalist (AB from Indiana University, 1968), I might not have bothered. When I became fairly sure that the Earth was hollow after reading various books on Earth science, I felt profoundly lonely and really separated from everyone else. Where is the reward for people to put themselves through that? I understand. So if you want to feel good (blissfully ignorant), don't read the rest of this book.

The hollow Earth theory is often mentioned in the same breath with the flat Earth theory, and the two are tossed out together with a chuckle and nothing more. Whatever you may think about these two theories, it should be clear to anyone that the hollow Earth theory deserves a lot more respect and consideration than the flat Earth theory. They are not in any way the same or equivalent. A serious case can be made for the hollow Earth theory.

When Isaac Asimov wrote in *The Robots of Dawn* that a thunderstorm's noise "rattled across a hollow world," he was fully aware that there were hollow-Earth believers out there in the public who could take this as a hint. I had sent him a hollow-Earth book of mine in which his possible hints had been discussed. Thousands of copies of Raymond Bernard's book, *The Hollow Earth*, had been sold. As an expert who was dedicated to opposing wild theories about UFOs and other science fictional topics, Asimov should have avoided the "hollow world" description like the plague. Instead, he used it, thereby throwing bait—red meat—to hollow-Earth believers. His willingness to do that should give us pause.

Chapter 4

How the Hollow-Earth Model Fits Facts and Prophecies

Earth scientists have left us hanging on some basic questions. Why do the north and south magnetic poles seem to be related to the axis without ever coinciding with the axis? And then there is the matter of our planet's rotation.

What holds the axis steady while the Earth's rotation rate is being nudged upward in response to solar storms? The answer is obvious, and yet...

Do scientists understand the Earth's rotation? I think they do, but they can't share their insight with us because the hollow planet theory is part of the explanation, and they aren't allowed to talk about that. I did a double take in 2022, listening to a report on National Public Radio in which various theories about Earth's rotation were being discussed. One of these was the idea that the power behind the Earth's spin was unleashed billions of years ago when the planet condensed into solid form, and it has been coasting ever since.

Seriously? How can it be coasting when the rotation rate receives a nudge from time to time in response to solar storms? The solar wind's powerful impact is directed—funneled—to the north and south magnetic poles after being deflected by the doughnut-shaped geomagnetic field. This powerful stream of charged particles is *always* impacting the magnetic poles, but its influence is only noticed when the solar wind gains strength as a result of solar storms.

It was John Gribbin and Stephen Plagemann, in their bestseller, *The Jupiter Effect*, who alerted the public to this influence of solar storms on Earth's rotation rate, along with resulting increases in seismic activity. The gravitational pull of Jupiter and other planets can shift the solar system's center of gravity to a point *outside* the sun, they wrote, and this shift can precipitate huge storms on the sun's surface. The fact that these storms ultimately result in increased earthquake and volcanic activity here on Earth was shocking to the public, but the full implications of Gribbin and Plagemann's work are far more shocking than that.

Solar wind impact is only one of many factors that can change Earth's rotation rate. There are seasonal variations caused by changes in the atmosphere and occasional changes unleashed by tectonic plate movements that alter the circumference of the planet. The solar wind, however, is the only one of these factors that is not cyclical in nature, meaning that the effects cancel each other out over time, and is the only factor which involves an outside force acting on the planet. Therefore, it is the only factor that can be said to be driving rotation, powering it, making it go.

Walter Sullivan in his book, *Continents in Motion*, also discusses abrupt changes in Earth's rotation rate, observing that they are remarkable considering that they involve the entire mass of the Earth. Neither he nor Gribbin and Plagemann made it clear that the observed changes generally amounted to increases in Earth's rotation rate. If scientists admit that Earth's rotation is powered by boosts from the sun, this principle raises a sticky problem.

Q: Why should scientists be reluctant to admit that Earth's rotation is powered by boosts from the sun?

RC: Because that raises the question, inevitably, as to what holds the axis steady while this solar wind push at the magnetic poles is taking place. If there were no force holding the axis steady, then the solar wind push would be expected to define an axis coinciding as precisely as possible with the location of the north and south magnetic poles. Obviously, we don't see that. Scientists are simply not allowed to explain the separation between the magnetic poles and the axis.

Q: But you claim to be able to explain that separation? Through the hollow planet theory?

RC: The axis cannot coincide with the magnetic poles because there is no land at the axis. Polar openings must exist, as Marshall Gardner observed, because orbiting matter cannot remain stationary at the axis. The outer shell of Earth is in orbit around the central planetary sun. This structure may sound bizarre and totally unfamiliar to the average person, but to scientists, it is not unfamiliar.

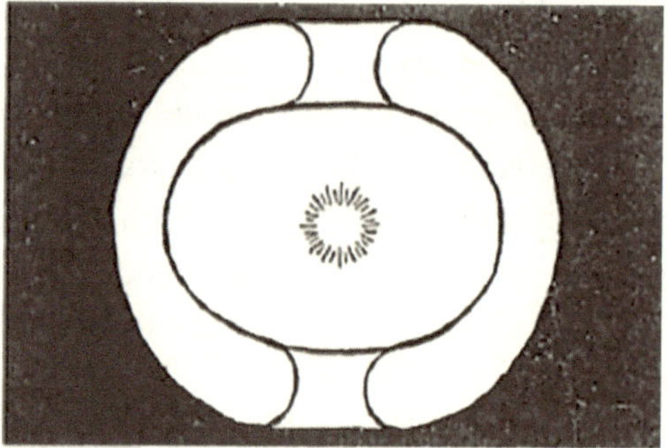

Figure 10. A cross section of a spheroidal planetary nebula, whose shape indicates what Earth must have looked like during its formation. This also would be the shape of a hollow planet.

Q: You seem awfully confident that scientists see things the same way you do.

RC: This is basic stuff. We shouldn't be arguing in the twenty-first century as to why the Earth is rotating. If there were no polar openings, the solar wind would simply define an axis at the magnetic poles, and its impact would not be far enough from the axis to drive rotation. The impact of the solar wind is not confined to a single point near each polar opening. Photographs of the polar auroras, which are set off by the solar wind, show that an aurora forms a *ring* around the axis, a ring which is thicker at one point, which must correspond to the location of the magnetic pole.

Q: You say the magnetic pole is a point, but then you suggest it is also a ring.

RC: In *The Hollow Earth* Raymond Bernard presented magnetic-field maps indicating that the magnetic poles are to some extent actually rings around the axis.

Q: Why don't the photographs of the auroral rings (see figure 9) show the central planetary sun shining through a polar opening?

RC: That light can be blocked by clouds. There are two "far UV" photographs of Earth in *Space Science and Astronomy* (Page) showing polar illumination on the nightside, which is identified as polar-aurora "caps." One of these "caps," in figure 133, is shown at an angle, revealing that it really looks like a solid cap.

Q: You said the aurora is a ring, not a solid cap.

RC: Exactly. The appearance of a "cap" of light must be created by the central planetary sun shining through a polar opening. A similar bright "cap" can be seen at the south pole of Mars in various photographs taken on August 4, 1969, by the Mariner 7 probe. (See figures 3 through 6.) Since Mars no longer has north and south magnetic poles, the "cap" of light cannot be identified as a polar aurora cap.

Q: That is odd that identical phenomena on two different planets would be identified differently. So how is the Mars south polar cap of light identified?

RC: It is simply called the polar cap or bright polar cap. It is identified specifically as ice in two Mariner 7 photos of Mars from *Pictorial Guide to the Planets* (Jackson), where the "ragged melting edge" is described in the caption. Ice, however, cannot shine on the nightside. And there was no proof when these photos were published that ice even existed on Mars. Only wet rocks have been detected near the poles of Mars as of now, not vast ice fields.

Q: So the scientists who identified the polar brightness on Mars as a polar ice cap were merely guessing, in your opinion?

RC: It appears they were guessing, yet their incorrect guesses have not been called out as a result of data that has been gathered in more-recent decades. I think the experts always knew that polar light on the nightside of Mars could not be accounted for by calling it ice because ice does not shine in the dark. I think they always knew the polar light was emanating from a central planetary sun, but they weren't allowed to say that. By presenting photographs that *prove* Mars has polar light on the nightside, they were actually defying the cover-up even though they lied about the source of that light.

Q: Can you prove that scientists knew the Mars polar light was on the nightside?

RC: If I can prove the polar light is on the nightside, I surmise that scientists who are smarter than I am cannot have failed to notice that.

Q: Isn't it more likely that your proof is flawed?

RC: Actually, a diagram that I did on this matter *was* flawed, slightly, yet it was good enough to prompt my congressman to submit my report to NASA for their comment. Representative Wayne T. Gilchrest had been a teacher of mathematics and so may have been intrigued by my attempt to use geometry to show how light bounced from the sun to the surface of Mars, creating a bright spot, and up to the Mariner 7 camera. My report was entitled, "New Angle Illuminates Mars Night-Side Light Mystery."

Q: Did NASA dispute your claim that the so-called polar cap was on the nightside, and did they explain the polar brightness?

RC: NASA did reply but had nothing to say about light on the nightside of Mars. If there was an easy explanation, such as, "The polar cap is on the dayside," or, "The brightness of the polar cap is caused by an aurora like we have on Earth," it seems they would have gladly provided that for Representative Gilchrest's sake if not for my sake. Members of Congress frequently address questions to government agencies.

Q: But no explanation at all was offered?

RC: This letter, dated August 17, 2001, bore the code "L:NKM:mtg:L/2001:486f." The opening sentence thanked Representative Gilchrest for submitting "information to support a theory that Mars is hollow with an internal sun." In other words, the entire inquiry was characterized as an attempt to promote a theory as opposed to a request for information. Even if NASA thought that of my work, it should have been self-evident that Representative Gilchrest was not promoting a theory himself but merely attempting to shed light on a mystery—the Mars nightside light mystery. Anyhow, the letter went on to say that NASA does not comment on theories originating with members of the public, and suggested that my work be submitted to a scientific journal. Various kinds of nonphotographic data about Mars were described at the end of the letter, and it was said that "all the data" supports the solid planet model. This was manifestly untrue, for photographic data is extremely important in astronomy, and the Mariner 7 photographs of August 4, 1969 strongly support the hollow-planet model, as do other Mars photographs.

Q: You admit, however, that your diagram was flawed.

RC: My diagram may have showed the polar brightness as being 98 percent on the nightside, whereas, in reality, it was only 92 or 95 percent on the nightside. With further study, I determined that the diagram was not as helpful as a simple sketch, which is provided in the first chapter, figure 1. With the image of photo C turned so the bright spot is halfway up on the right side, it is clear that the dayside can only extend for 180 degrees around the edge of the image, with the bright spot of reflected light falling halfway along that 180-degree span. This is a planet in half phase. The polar "cap" is well beyond that 180-degree limit and clearly facing away from the sun's light. Photo C has been extensively doctored to give the appearance that illumination is extending much farther to the left than shown in the sketch. Yet the location of the bright spot of reflected light tells us that that area on the lower left must be in shadow, which means that the polar brightness is unexplained. Photos A through D were all altered to obscure the day-night border where it approaches the right side of the polar "cap."

Q: Most people don't spend a lot of time worrying about the lighting of Mars.

RC: Do you think people would be comfortable knowing a hollow planet is winding down in our vicinity, losing rotation speed, and rapidly approaching the point where the shell will not be spinning fast enough to remain in orbit around the central star?

Q: So you think Mars will fall apart?

RC: Yes, because it is not receiving the solar wind's push at north and south magnetic poles as we see on Earth. Mars no longer has north and south magnetic poles, and so the solar wind cannot be concentrated at the key points where its pressure would provide a boost to rotation.

Q: How do you know Mars ever had north and south magnetic poles in the past?

RC: I have heard scientists state clearly that Mars at one time had north and south magnetic poles but then lost them. Evidently, they conclude that those poles had to exist in past ages, or Mars would not be rotating today. Also, they may deduce that those poles used to exist because of the existence of old lake beds and riverbeds, which would not have contained water without a protective north-south magnetic field to deflect the solar wind away from most of the surface. The loss of the protective geomagnetic field would have meant the loss of most of the planet's atmosphere and water.

Q: So you think scientists are concerned about the lack of solar wind support for the rotation of Mars. You think they are worried about Mars falling to pieces, but they aren't allowed to speak up about it?

RC: Yes, that is what I think, and I can prove they have hinted about the Mars threat as well as hinting about a coming axis shift and about the breaking of brains to enforce the cover-up, and I can provide coded messages from them to support my conclusions.

Q: People are not going to care about that kind of evidence because the hollow planet theory is so completely weird and theoretically questionable.

RC: The hollow planet theory not only explains how the axis and magnetic poles can remain separate, enabling rotation to be powered. It also explains how the axis tilt is reestablished in case of a shift in the axis. When the geomagnetic field reverses itself, the north and south magnetic poles can wind up not only in opposite hemispheres but in totally new locations. The solar wind's pressure at those new locations would have to shift the axis, and one would then have to explain why the axis tilt is adjusted, or why there is an axis tilt at all.

Q: In trying to justify the hollow planet theory, you find yourself introducing another strange and controversial concept, the shifting of Earth's axis, which hardly strengthens your case.

RC: I didn't make this up myself. The Earth's magnetic field does reverse itself, and it is our bad luck to live in a time when that rare event is approaching. Various experts have said so, and a *Scientific American* article of December 1989 spells out the details (Bloxam and Gubbins). The axis-shift predictions Edgar Cayce gave during trance readings were investigated by Jess Stearn for his book on Cayce. A geologist who asked to remain anonymous told Stearn that it was true that an axis shift would be expected as a result of a geomagnetic field reversal, for which we are "overdue."

Q: Assuming that he was telling the truth, how do you think the axis tilt would be preserved through the upheavals involved in an axis shift?

RC: The key here is a passage from the *Encyclopedia Britannica* (1974). I looked up information on the planetary nebula, which, according to an early theorist, provides a model for the structure of a hollow planet. Marshall B. Gardner thought a planetary nebula, with its central star and orbiting shell, could be a stage in the evolution of a hollow planet. He didn't realize that the planetary nebula is much too large to become a planet and is not part of a solar system like Earth and Mars, those being the planets Gardner considered to be hollow. The structure of the nebula, however, provides a clue regarding the axis tilt on Earth. Britannica states that the matter orbiting the central star of the nebula can take the form of spirals, rings or helices, or shells that are spheroidal or truncated. It was found early in the twentieth century "that oblate, spheroidal or truncated shells, thinner at the equator than at the poles, accounted satisfactorily for the appearance of many planetaries." This tells us that the temperate regions of a hollow Earth would be thicker and more massive than the equatorial regions, and so the sun's gravity would attract the temperate regions alternately, as opposed to attracting the equatorial regions.

Q: So the axis shift would cause the new temperate regions to thicken up, causing the axis tilt to be adjusted?

RC: Yes, and this would produce seismic activity. The equatorial bulge would move to the region of the new equator. Once again, there would be seismic activity, including the rupture of super-volcano magma chambers. Edgar Cayce predicted that the coming axis shift would be associated with many volcanic eruptions in the "Torrid area." The reshaping of the equatorial areas, new and old, makes this prediction easy to understand. The polar openings are also rotational effects, and they would be reshaped to conform to the new axis. Some lands would be obliterated or plunged into the deep freeze in the process while others would become warmer, and Cayce did predict widespread climatic reversals. The changes in the equatorial bulge would bring about instant changes in the levels of the ocean in various places, swamping some lands and receding from other coasts to reveal new lands.

Q: And all this was predicted by Cayce?

RC: Actually, yes, it was. It was a voice speaking through Cayce while he was in a trance. The voice would provide health advice and various predictions for individuals, and then there would be bits and pieces of cataclysmic information given in various readings.

Q: It sounds about as reliable as going to a fortune teller.

RC: It was up to us to collect the cataclysmic predictions and evaluate them. Stearn did the collecting, and his consultant, the anonymous geologist, commented that the prophecies were plausible. His most significant comment was that the axis shift prophecy could be fulfilled by a geomagnetic field reversal, for which we are "overdue."

Q: Were the other prophecies plausible?

RC: Taken one by one, they cannot be evaluated very well. This is because each earth-change prediction makes sense only if one knows where the new axis will be located. Taken together, however, these various predictions form a consistent picture around the idea of a new pole of rotation in Northern Europe. Such a consistent picture could not be formed by your average fortune teller making wild guesses.

"The upper portion of Europe will be changed as in the twinkling of an eye." This was the prediction that to me indicated the formation of a new pole where Northern Europe is now. The sudden change would be brought about by the migration of the present north polar opening to the location of what would become the new south pole.

Q: You are assuming the Cayce prophecies include the hollow-Earth concept.

RC: It's pretty obvious. The polar shift prediction was dated for the year 2000. Also, the forty-year period ending in 1998 "will be hailed as the time when His Light will be seen again in the clouds." The light in the clouds must refer to the inner planetary sun shining up into the sky through a new polar opening. It would be making an appearance to people in a way that now does not occur because people don't live close enough to the pole to see that light in the sky.

Q: So the polar light would appear to many people who were living more or less in the path of disaster?

RC: Yes. Even though Cayce's dates were not accurate (and they were probably made too early on purpose to give us more lead time to understand), the important thing is that the light show in the clouds is dated for about the same time as the axis shift, and so it would be a manifestation of the axis shift.

Q: Do you see the Cayce prophecy about "His Light" coming in the clouds as relating to the Second Coming of Christ?

RC: I think it was intended that way, since Christ is the one who is known for his promise to return in the clouds.

Q: But does that mean that Christ himself intended his Second Coming prophecy to relate to the appearance of the central planetary sun in the clouds?

RC: That point can be argued based on Holy Grail symbolism, and we will look at that. But it is difficult to be certain exactly what his meaning was.

Q: Well, the Cayce prophecy on axis-shifting would have been clearer if it had just said the inner sun's light will shine through polar openings.

RC: The truth was evidently not allowed to be told that clearly. If we were going to save ourselves from this disaster, we evidently were required to *work* to acquire the necessary understanding. It appears there has been a kind of game going on to determine whether we are worthy to survive or whether we are too stupid and foolish, and there may be an issue of racial inferiority.

Q: So the people in charge, ruling from hell or wherever, may consider themselves to be a superior race?

RC: They may be a different race or even a slightly different species. Anyhow, we have been allowed to have some clues to the truth, but not the whole truth, for whatever reason. One big clue is the way the Cayce prophecies fit the picture of an axis change involving a new pole in what is now Northern Europe. A warmer climate is predicted for Alaska, which would be one obvious result of a pole being relocated to Northern Europe. Also, with that new pole, the new equator would be much closer to Japan, and so the new equatorial bulge would bring high water to that area. Cayce predicted "the greater portion of Japan must go into the sea." Cayce predicted that Russia would be hailed as "the hope of the world." With a new pole in upper Europe, eastern Russia would become warmer than before and therefore able to accept refugees from Japan and other threatened areas. The eastern United States would become farther from the equator and therefore would see less of the high water associated with the equatorial bulge. For that area, Cayce predicted that new land would appear off the coast.

Q: The prophecies do seem to point to the formation of a new pole in Northern Europe, but the big question, then, is why. Why is most of Europe slated for destruction or the deep freeze?

RC: I think it has something to do with the distribution of lands and seas across the globe. Ursula B. Marvin observed in her book, *Continental Drift: Evolution of a Concept*, that all land masses have moved north in recent geologic time. Obviously, this cannot go on indefinitely. There was a beginning to this trend, and there will be an end to it. If land has been seeking what is now the north pole for many ages, the time may have come for the north pole to be relocated far from land, in the southern Pacific Ocean.

Q: It is unfortunate that Cayce did not receive an answer to this question as to why the poles are to shift in a particular way. It is impressive that someone had the technological skill to project messages to us using Cayce as a medium. But still, the individual's identity is unknown. If these predictions are to be credible, they must be confirmed by our own scientists.

RC: I can provide hints, riddles, and coded messages that confirm the picture provided by Cayce, mainly taken from the work of Isaac Asimov and Carl Sagan. Some of these covert signs are based on Grail symbolism that extends back in time for many centuries. I believe the symbolism can be understood, and the totality of it will give us a powerful sense of the truth striving to be set free.

Chapter 5

A History of Grail Traditions That Point to a Hollow Earth and Coming Shift in Earth's Axis

In a way, it doesn't matter whether Jesus or other ancient teachers gave hints and symbols pointing to a hollow Earth or whether Hitler's backers in the Thule Group believed in a hollow Earth and in mythical continents that had been destroyed in an axis shift. The reason it doesn't matter is that we don't get our information about the world from ancient religious leaders or occult secret societies.

On the other hand, if we skip over the history of hollow-Earth belief and concentrate only on the covert messages of Isaac Asimov and Carl Sagan, the two of them will look like anomalies—complete flukes—not to mention how peculiar I might look for building a grand case on covert messages from two men.

One of my favorite hollow-planet hints is from Egyptian mythology, as reported in Flavia Anderson's book *The Ancient Secret.* She presents a picture of Ra, a son of the sun god, shooting arrows representing light beams at a polar constellation known as the Thigh, or the Haunch of the Bull. (No doubt many of us would call it the Little Dipper today.) She makes a connection between the constellation and the thigh wounds which are part of many Grail legends. The real message in the picture, however, appears to be that a junior sun sends light beams toward the North Star and a few other stars, because those are the only stars that can be seen from within the hollow-Earth shell where that sun is located (unless one considers the south polar stars). In other words, this picture places the junior sun at the center of the Earth looking through a polar hole.

Anderson also says the Hebrew Jacob must have been considered a solar hero because he was the father of twelve sons, who correspond neatly to the months of the year (or the signs of the zodiac). Similarly, in *Mysteries of the Holy Grail*, Corinne Heline states that King Arthur mystically represented the sun while the twelve knights of the Round Table represented the signs of the zodiac. These two scholars on the Holy Grail implicitly must have thought that Christ was a solar figure because he had twelve disciples who represented the signs of the zodiac.

One might ask why a leader surrounded by signs of the zodiac should be seen as a solar figure. The image of a sun surrounded by zodiac signs, which are earthly inventions visible only from our perspective, must be telling us that there is a sun at the point on which the zodiac signs are centered, that is, at the center of the Earth. The central leader is a solar figure simply because a hollow-planet message is being constructed here for those who can figure out the diagram.

The Round Table of King Arthur represents the Earth, according to Heline. There are twelve (zodiacal) figures seated there—Arthur and only eleven of the knights—with an empty seat that would have been occupied by a knight who was disgraced and not present. Similarly, at the Last Supper, there were only eleven disciples at the table with Christ, making twelve (zodiacal) figures altogether after the departure of Judas. Heline even places a holy chalice at the center of the Earth-table, containing the blood of Christ, which on Good Friday glows with a sunlike brilliance. The hollow-Earth concept is mentioned in her book when she says a certain mountain is reputed to be hollow, with channels leading to the center of the Earth.

Whether Heline's versions of Grail legends is accurate is not as important as the fact that she was seizing every opportunity to present hollow-Earth imagery to the reader. *She* felt a hollow-planet message was indicated, and she sought to amplify that.

The words of Christ and events in his life can sometimes be seen to contain hollow-planet hints. He said: "You are the light of the world. One does not light a candle and place it under a bushel, but on a stand." With these words, he conjures an image of the light of the world surrounded and hidden by a bushel basket—an image closely resembling the hollow-planet model shown in figure 10.

Christ's birth is sometimes said to have occurred in a cave that was used as a stable. This suggests that he represents something located in a hollow within the Earth. His cross has been said by mythologist, Joseph Campbell, to represent the world axis. If he is viewed as a solar figure, the message is that a solar figure or sun is *united* with Earth's axis. Campbell states that in mythology the "world mountain" represents the world axis. If Christ considered this to be so, then his statement that faith can move a mountain paints a picture of a world axis shift. In view of the hollow-planet symbolism in his life and teaching, we can surmise that his promise to return in the clouds at a time of cataclysm was a symbolic statement about the central planetary sun's appearance in the clouds near new polar openings after an axis shift. We can't be sure of that, and we can't know what kind of real information he had to back up his hollow-Earth beliefs. But I feel it is fairly clear that he laid those beliefs out for us in an extremely dramatic manner.

We should add that his end-time parable of the wise and foolish virgins appears to be a portrait of a solar system in which all planets shine like lamps, but not all are lit from within a hollow enclosure. The virgin was a commonly used symbol for a planet in ancient times. The large number of virgin-planets in the parable suggests that Christ somehow knew of the existence of more planets than most people knew about in ancient times. In other words, his prophecies were based not on superstitious belief but on authentic, advanced knowledge.

The book of Revelation, chapter 12, contains an apparent description of an Earth axis shift, described as a shifting of the stars in the heavens. A woman clothed with the sun and with the moon at her feet and a crown of twelve (zodiacal) stars on her head (she must represent Earth) is menaced by a dragon whose tail sweeps a third of the stars from the heavens. The dragon likely refers to a constellation known as the Dragon. I believe it was Helen P. Blavatsky who explained that a star shift, or axis shift, is traditionally described as a sweeping of stars from the heavens by the Dragon constellation, whose tail sweeps in a new way during the star shift. Blavatsky mentions axis shifts, or pralayas, frequently in her book, *The Secret Doctrine*. It is difficult to prove that the biblical description of a star shift was intended to be a description of an Earth axis shift. However, various esoteric traditions link star shifting with cataclysmic events.

Anderson describes prophecies by the "heathen Flegetanis," who predicted changes in the courses of the stars along with planetwide upheavals. In *The Impending Golden Age* by John Lowell and Kaspar Nureddin of the Sanctilean Syntonium (a branch of freemasonry), we read: "Today our North Pole points approximately at a star called the North Star. Following the tilting of the axis of the Earth, our North Pole will no longer point at this star. The star will not have moved. The axis of the Earth will have tilted to cause the star to appear to have moved."

Also, they write: "This tilting is popularly called the Star Shift. The shifting of the axis of our Earth results in what appears to be the shifting of all the stars in the heavens." In this process, of course, some stars would drop out of sight permanently while others that had previously not been seen in one's hemisphere would start to be visible.

Mme. Blavatsky quotes Plato's description of changes in the courses of heavenly bodies in connection with the supervision of mankind by higher beings (p. 372):

> According to Plato, in order to obtain clear
> and precise ideas on royalty, its origin and power,
> one has to turn back to the first principles of history and tradition. Great changes, he says, have
> occurred in days of old, *in heaven and on Earth*,
> and the present state of things is one of the results
> (Karma). Our traditions tell us of many marvels,
> of changes that have taken place in the course of

> the sun, or Saturn's reign, and of a thousand other
> matters that remain scattered about in human
> memory, *but one never hears anything of the EVIL*
> which has produced those revolutions, nor of the
> evil which directly followed them. Yet… that Evil
> is the principle one has to talk about, to be able
> to treat of royalty and the origin of power.

This indicates that changes in the courses of heavenly bodies, specifically the course of the sun, can be brought about by royal rulers of our planet, according to Plato. Lowell and Nureddin also describe the supervision of humanity by higher beings following major cataclysms.

Louis Pauwels and Jacques Bergier report in *The Morning of the Magicians* that Adolf Hitler believed Earth to be hollow and also believed we are living in the *interior*. This makes Hitler's hollow-Earth belief appear to be poorly grounded in evidence or any semblance of careful thought. In *The Occult and the Third Reich* by Jean-Michel Angebert, it is reported that Hitler had stated, "the structure of Europe" will be "capsized in an immense cataclysm." Here again, it is not clear that Hitler had any understanding of what the scientific principles behind such an event might be. He may have been simply repeating something that was told to him by members of the Thule Group, who supported him behind the scenes. The Thule Group members believed the northern continent of Thule had been destroyed in an axis shift. The Thule Group was said to have a liaison with the Green Dragon Society in Japan. The very fact that the term Axis Powers was used to refer to the German-Japanese alliance suggests that these were two countries that anticipated devastation for their homelands as a result of an axis shift. The Cayce prophecies certainly are not favorable for either Germany or Japan. Perhaps, somehow, their leaders could see a grim future ahead, which drove them to try to acquire land elsewhere.

One year after I sent Isaac Asimov my 1981 book, *Hollow Earth Mysteries and the Polar Shift*, Asimov's novel *Foundation's Edge* contained an echo of Holy Grail symbolism, which I had discussed in my book. What I wrote in 1981 was as follows:

> In *Mysteries of the Holy Grail* Corinne Heline states that King Arthur, surrounded by the twelve Knights of the Round Table, is said to represent symbolically the sun surrounded by the twelve signs of the zodiac. Christ, surrounded by the twelve apostles, would be in the same symbolic position. How do we know that the sun in this symbolic description is inside the Earth? It is quite simple. The signs of the zodiac exist only from the perspective of the Earth. Seen from other points in the universe, the constellations would be arranged differently. Even though the changes might be slight at slight distances from the Earth, it is clear that symbolically any object, such as a sun, placed at the center of the signs of the zodiac must be represented as an object on— or inside—the Earth.

What Asimov did in *Foundation's Edge* was to create a table of twelve with an empty seat—empty until its proper occupant shows up late to take his seat. This echoes Heline's discussion of the Round Table, with an empty seat because of the absence of one of the knights, all this being a recreation of Christ's Last Supper, with an empty seat created by the departure of Judas. Even though my book had not mentioned the table of twelve as described by Heline, the hollow-planet symbolism in the passage I have quoted was similar to the hollow-planet symbolism in King Arthur's Round Table, which, according to Heline, represented the Earth. A chalice at the center of the table contained Christ's blood, which was said on Good Friday to shine brilliantly like a sun. One sign that Asimov intended his table of twelve to be related to the Grail table representing a hollow Earth is the fact that the character who claimed the empty seat has a name meaning "star." His name is Stor Gendibal, and stor is an archaic form of the word star. This suggests that the twelve figures at the table are stellar, or zodiacal, figures.

In his 1986 novel, *Foundation and Earth*, Asimov presents a planetary image surrounded by twelve (zodiacal) entry stations, and hovering over the center of the image—the southern pole of the planet—is a spaceship called the *Far Star*. An "empty seat" is created when one entry station leaves its position and approaches the *Far Star*. So Asimov here created an image of an actual planet with zodiacal figures and a central solar symbol, and even an "empty seat," all corresponding to the Grail table as described by Heline. This seemed to be his way of saying: "Yes, the Grail table does represent a hollow Earth."

There is a strong hollow-planet symbol in *Foundation's Edge* in the form of the "mayor" of the planet Terminus, Harla Branno. She is suspected of being "a hollow shell with a Second Foundation (enemy force) content." As the chief political figure on her planet, she undeniably represents a planet in a political sense, which makes it easy for us to see her as a symbol of the planet itself. Describing a planetary symbol as possibly being a hollow shell suggests that a planet itself can be hollow. Her first name, Harla, even sounds a lot like "hollow." Her jaw is like granite, and her hair is iron gray. These descriptions associate her with the rocks and minerals of a planet.

All this from Asimov is in the realm of hints. And don't forget his phrase "rattled across a hollow world" in *The Robots of Dawn*. I wrote of his hints again and again and again, then much later discovered the SOS signs, and the riddles, and even coded messages adjacent to punctuation. He spent his entire career planting the seeds of understanding for us to discover later. It can be painful to look at all this, to perceive what he went through to try to get through to all of us. For that matter, look at what Christ went through to convey his cataclysmic signs in a way that would be noticed thousands of years later.

Chapter 6

Thirty Years to Civilization's
End—Asimov's Prediction

My discovery of Isaac Asimov as a secret advocate of the hollow Earth theory came as a result of the interview of him that appeared in *Mother Earth News*, September–October 1980. The front cover asked, "Thirty Years to Armageddon?" This headline was based on Asimov's prediction that civilization, or civilization as we know it, was unlikely to survive to 2010. His prediction that future space pioneers would "hollow out the mini-planets" appeared in large print. By mini-planets, he was referring to asteroids. The language used, however, called to mind the idea that planets could be hollow.

There were two other possible hollow-planet hints in his remarks. He said, "we are used to living on the outside of the hull" of Earth, if it is viewed as a spaceship. Also, he said future space pioneers would be living "*inside* a world." But what he described was merely the inside world of the space traveler. His pictures showed him looking somber.

At the time of reading all this, I was exploring the idea that an axis shift on a hollow Earth could fulfill the prophecies of Edgar Cayce, especially as to the appearance of a special light in the clouds at the time of an axis shift. The formation of new polar openings in Northern Europe and the southern Pacific Ocean, which I was beginning to anticipate, would unleash unimaginable horrors that would

39

indeed threaten civilization. With this picture in my mind, reading his possible hollow-planet hints along with the prediction of civilization's end, it seemed Asimov had the same picture in mind that I had. It seemed his hints were a way of taunting people about cataclysmic knowledge he had which he was unwilling to share. Suddenly, however, I saw him not as deliberately withholding knowledge but as desperately trying to reveal secrets that somehow could not be told. I burst into tears and wept for a long time.

This idea that scientists could not tell the whole truth about hollow Earth and cataclysm was reinforced when I read Carl Sagan's book *Broca's Brain*, whose title sounded like, "broke his brain." I began looking for hints in the works of both these scientific experts. I also read a number of scientific texts and some UFO literature, which I thought might shed light on the cover-up.

There was an urgency to my research and writing in the 1980s and 1990s because of the deadlines that had been set by Asimov and earlier by Edgar Cayce. The year 2000 went by without the axis shift that Cayce had predicted. This resulted in a loss of public interest in his prophecies. He came to be viewed as just another alarmist picking a date for the end of the world and being totally mistaken. People who dismissed Cayce so easily probably never examined the prophecies in detail to see how they take into account the various ways the planet would be reshaped in an axis shift.

Instead of dismissing Cayce's predictions, I began to think the dates may have been deliberately set too early to spur someone like me to begin working hard at an investigation that was bound to take a long, long time. If the shift really had been bound to occur in the year 2000, most of my best evidence would not have been discovered in time to do any good.

With Asimov's deadline of 2010, once again I felt under pressure as it approached. I did a long book at a publisher's request around 2005. This was for Global Insights. He didn't use it after all, but at least he did return it. This book described the threat of a Mars collapse, which I had just concluded was a much bigger problem than a coming Earth axis shift. I continued to do occasional newsletters beginning in the 1990s for the next three decades.

Asimov's deadline came and went. Perhaps he, too, had picked an early date for the end of civilization to spur possible scholars to keep working hard. If 2010 was not the real date for an immense cataclysm, what date can be found in his work that might tell us what the real deadline is?

The word seventy occurs in one of the made-up terms that are spread throughout Asimov's stories featuring the demon Azazel, who comes from a world where a different language is spoken. These made-up terms allowed Asimov more flexibility than ordinary names as far as supplying him with exactly the letters needed for a particular riddle. In his story "It's a Job" (43), published in 1991, Azazel uses the term "seventy dworshaks."

The first thing we should take note of is that we can take "seventy years" from this riddle. Seventy years from 1991 is 2061. If we knew an immense catastrophe would take place then, this would shape our decisions about what big projects are worth undertaking and which ones should be set aside. It would also help us determine which wars are worth fighting.

The word shake is suggested by "shak" in this riddle. The letters for "Earth" are present. Here is my solution: "Eye worry over events that are to shake or even destroy the Earth seventy years down the road."

A riddle containing "axis shift," as well as a dated prediction for when it may be expected to occur, would be helpful. Carl Sagan appears to have provided us with this kind of a riddle in his novel *Contact*. He sets the stage by having aliens from Vega make remote contact with our scientists who have been scanning the skies for possible messages from beyond. News of this contact results in a lot of public hysteria, including fears of doomsday and even an Earth axis shift as the year 2000 approaches. The aliens have sent plans for a machine whose function is unknown, but which has room for five passengers.

Stefan Alexeivich Baruda gives a fiery speech, stoking the doomsday fears that already exist. Though he does not specifically mention an axis shift, the fears in the public mind already include the idea that this alien contact may precipitate an Earth axis shift.

His name *contains the letters for "axis shift,"* which I must tell you is an extremely unusual occurrence in a character name. I can't think of any other character name that contains the letters for "axis shift."

This is not an easy riddle. Several of Sagan's easier riddles will be seen later. After "axis shift" has been taken from the name, however, the riddle becomes shorter and easier to decipher. After some failed attempts with this riddle, I noticed that the letters for "field reversal" are also present. And so it appears that Sagan wants to tell us that a (geomagnetic) field reversal is going to cause or set off an axis shift. This is the same information that Jess Stearn received from an anonymous geologist regarding Edgar Cayce's axis-shift prophecy, as mentioned earlier.

The letters for "decades" can be found in Baruda's name. Unfortunately, there is no definite number of decades given in the message. I included my name in solving this riddle because I did send Sagan more than one of my books describing how a field reversal would set off an axis shift. Here is my solution to Stefan Alexeivich Baruda: "A field reversal shall cause Earth's axis tu be shifted (as Ruth Anne Leedi rites) in five, seven, nine decades." Seven decades from 1985, when Sagan's book was published, would put the axis shift in 2055, which is fairly close to Asimov's prediction in the "seventy dworshaks" riddle. We might expect their predictions to be coordinated, based on the fact that they both did riddles and at least one SOS sign, and their riddles followed the same general themes.

The danger of a collapse of slow-rotating Mars, which is described in various riddles, raises the big question as to when that existential threat is likely to materialize. A few riddles about the Mars threat are as follows:

"Mars is not yet (totally) limp," Limmar Ponyets (26).

"Mars is about to go limp, to vibrat, crumpl, collaps," Alistair Tobago Crump, VI (11).

"Mars end nears," Mandress (29).

"Mars is doomed. I am scared." This is from Mordecai Sims (19), in a story where Earth's lifetime is shortened.

"Mars shall end and its inner star shall threaten the Earth," the Marshal Nedelin (126).

"Do not let the Mars star land on Earth," Leander Thomson (17). A large orange ball lands in the wrong places in this story.

To find dated predictions about the collapse of Mars, we need riddles that contain the word years, as well as Mars, and perhaps also doomed and Earth and threaten. I see four riddles that may help us out. Two of these come from *Nemesis*, where Asimov portrays a secret star heading for Earth. Another comes from *Foundation's Edge*, where Asimov gives a long riddle containing the letters for "Azimov" all grouped together. This was the original spelling of what is now the name Asimov, as the author explains prior to his story, "Spell My Name with an S" (76). It is *that* story which contains the name Marshall Sebatinsky, where Asimov gives us "Mars" and "shall" in a very obvious way. In that name, we can find: "Mars shall break in thirty, siksty, ninety years. Its inner star then shall threaten the Earth." If Sebatinsky's wife, Sophie, is included in the riddle, the solution may be: "Mars rotation shall stall. Its hollo shell shall break apart in perhaps thirty, siksty, ninety years. Its inner star then shall threaten the Earth."

In Endomandiovizamarondeyaso (29) we can find: "I, Asimov, say Mars is doomed and may end in seven-zero or nine-zero years." If we do not use the word zero here, we have to spell Asimov as Azimov in order to use the Z. Even though the riddle seems to encourage us to do this, it is not permitted in riddles to misspell anything if it can be spelled correctly. So I think the phrase "in seven-zero or nine-zero years" is intended to be used.

Turning to *Nemesis*, the heroine is Marlene Insigna Fisher, in whose name we can find "Mars," but little else that we need for a prediction. She is closely bonded with the moon-world, Erythro. Adding Erythro to her name, the riddle may be saying, "Mars rotation is stalling. Its hollo shell is going to shatter in forty, fifty, eighty, ninety years. Its inner star shall then threaten the Earth."

Another character in *Nemesis* is closely associated with Erythro. Siever Genarr is commander of the Erythro Dome. In "Siever Genarr, Erythro Dome," we can find: "Mars is doomed, and its inner star is on the verge ov destroying the Earth in thirty, seventy, eighty, ninety years."

These four riddles date from 1957, 1982, 1990, and 1990 respectively. Therefore, these riddles generate the following possible deadlines as far as the collapse of Mars is concerned: (1) 1987, 2017, 2047; (2) 2052, 2072; (3) 2030, 2040, 2070; and (4) 2020, 2060, 2070, 2080.

The possible predictions seem to fall mainly between 2040 and 2070. Considering how difficult it must be to estimate when Mars might collapse, this may be as close as Asimov intended to come to pinpointing an exact date. I have not yet found any prediction from Carl Sagan as to when Mars might collapse, but the "Marshal Nedelin" riddle from his novel, where we easily find, "Mars shall end," shows his awareness of the problem.

It appears there are two possible remedies for the Mars threat, as we shall see.

Chapter 7

SOS Signs in "The Hollow Earth," an Isaac Asimov Essay; "I Need Your Help. Please Decipher and Share Riddles."

Anyone can need help at times, even very intelligent science writers. The use of SOS signs by Isaac Asimov and Carl Sagan shows they were looking for help in revealing hidden truths to the public. Someone trying to end a cover-up regarding coming cataclysmic events is tackling a huge task. *Of course*, they needed help.

SOS signs are included as part of the plot in Asimov's story "Old-Fashioned" (82). I couldn't help noticing that sets of dots occurred frequently in the story, with two such sets being paired together in three different places, The S in Morse code consists of three dots. Where four dots occurred together, it was clear that one was a sentence-ending period, and the other three, taken together, signified a pause. And so what I was seeing was six different S signs grouped in sets of two in three different places. In between the two S signs, in each case, was at least one word indicating a circle or letter O. In one case, the word between the two S signs was "orbit." In another case, the O-word was "hole," and in the final case, it was "pill," as well as the word "oh." Toward the end of the story, there were many dashes, and we recall that in Morse code the letter O is made using three dashes.

As to the purpose of these possible SOS signs, I think Asimov wanted people to take seriously the story plot, especially the part about a small black hole that could potentially destroy the Earth. This small black hole was able to hold a spaceship in orbit around it. In other words, it was behaving like a sun. I think the message was that a small sun had the potential to devastate or destroy the Earth. If Asimov had thought the situation was impossible for us to handle, he wouldn't have been anxious to draw our attention to it. I think he was hoping we could divert the Mars star and its orbiting shell onto a new path before the planet self-destructed, and he wanted us to apply ourselves to that goal. He may have seen a possible threat from Venus as well, but that would have been more remote.

"Hollow Earth" Essay Ends in "SOS"

The third hollow-Earth book I sent to Asimov was entitled *Hollow Earth Apocalypse: Asimov's Warning* (101). This inspired him to write an essay entitled "The Hollow Earth" (55). He noted that I had quoted passages from his work, taken out of context, and interpreted them as supporting the hollow Earth theory. In other words, he was saying that the author (not named) had accused him of *hinting* that the Earth was hollow. The greatest possible hint he could then drop would be to denounce the hollow Earth theory and then turn around and end the essay with an SOS sign that would negate everything he had just said. And that is exactly what he did.

The final sentence was, "So say I!" The first half of this six-letter sentence is "Sos." The sentence can easily be converted to "SOS, say I," with the addition of another S. The exclamation point, if rotated, consists of a dot and a dash, the basic elements of Morse code. Immediately prior to this peculiar final sentence, there are two unusual figures of speech. He said the hollow Earth theory is a "potpourri of poopery" and an "inanity of insanity." These are both modified ABA forms in which a single letter causes the third part of the expression to sound slightly different from the first part. The SOS sign, of course, is an ABA form in which the third part is identical to the first part. With the first expression, the letter O makes the difference between the "popery" sound and the sound of "poopery." The letter converting "inanity" to "insanity" is an S. So these expressions feature the letters of the SOS sign as the pivotal letters. The middle word of each expression is "of," which can be abbreviated as "o'," just as O is the middle character in the SOS sign.

Turning to the beginning of the essay, I noted that the words "astonishing" and "socks," both containing S, O, and S in that order, are used together in the sentence describing his first reaction to my book. He said that just when he thought he had seen everything and nothing could break life's round of dullness, "something astonishing comes along and socks me right in the funny bone." Later he said "aspersions" were being cast on his competence by my book. This word also contains S, O, and S in that order. It was a bit of a stretch for him to use that word, since aspersions are deliberate criticisms, whereas I had praised him highly for his hollow-planet hints.

The significance of words containing S, O, and S in order can be appreciated by checking Asimov's poems "I Just Make Them Up, See!" (74) and "The Author's Ordeal" (84). These poems include a total of three words containing S, O, and S in order. These words are "so's" and "socks" and "spots." The "so's" is a contraction of "so" and "as" (who does that?) and is an extremely rarely used word. Its resemblance to the SOS sign is striking.

Charles Soskind is a name where it looks as though Asimov might be sending an SOS sign. The message here appears to be: "Each riddle I do is an SOS sine. I seek aid." Soskind does seek aid in the story (21).

Sossafer Chasids is a term beginning with "Sos," which can be found in Sagan's novel. In this term, Sagan may have been saying: "I fear if I *said*, 'SOS,' I cood die, so I choose a safer SOS." The safer SOS is found in the word Sossafer.

In the section of *Contact* featuring Ian Broderick, whose name can spell "broken brain," Sagan has placed many sets of three dots, indicating pauses in a phone conversation. If one looks between these possible Morse code S signs, there are two intervals where three dashes or hyphens can be found, possibly indicating O in Morse code. In *one of these intervals*, there is a reference to dots and dashes, which are the basic elements of Morse code. It is clarified later that the transmission from space consists *not* of dots and dashes but of the ones and zeroes of computation in base 2. It would, after all, be absurd to suppose that a message from another planet would be presented through the dots and dashes of Morse code, a code the aliens had never been exposed to. The Morse-code reference was not exactly a mistake. It was evidently Sagan's way of confirming that the triple dots and triple dashes (or hyphens) were *intended* by him to be seen as an SOS sign.

End-of-World Riddles Being Ignored

A call for help by Asimov can be seen, possibly in the name Nellie Griswold (53), where we can find: "End o' world riddles need reeding. No one is willing." Evidently, Asimov knew of some puzzle-masters who were able to interpret his riddles and other hints, and he wanted someone to volunteer to decipher his riddles and share their meaning. (This message predates my work on Asimov's hints.) Another call for help along this line can be seen in Brinsley Sheridan Cooper (92), where we find this message: "Brain is disabled. I need your help. Please decipher and share riddles." Similarly, in Sindra Lambid (66), we see: "Mi brain is disabld."

Carl Sagan defines his brain limitation in a specific way in Billy Horstman (126): "My brain is not abl to say, 'Mars is hollo.'" There is a suggestion of bondage here, since Billy's father faces prison time. Sagan's easiest riddle is the name of a prisoner—Tyrone Free.

When we see that a prisoner has the last name Free, our natural reaction is to think, "But he is *not* free." Then we notice that Tyrone contains the letters for "not." Now all we have to do to complete the message is to use the letters Y, R, and E. My solution is: "Eye R not free." Adding the name of Abonneba Eda, who visits Tyrone Free in prison, the riddle message may be: "Eye be not free. Brane be bent, torn and abraded." The letters for "brain" occur together at the beginning of Ian Broderick (126), where we find: "Broca died, brain broken." This told me I was correct in saying, beginning in 1981, that Sagan's book title *Broca's Brain* could be hinting at the message, "Broke his brain."

Adding the name of Broderick's countryman, Annunziata, to the riddle, we can find: "U can C Broca's brain. It iz bent, torn and broken." You can examine the damage to Broca's brain because it is preserved in a jar.

"Scientists' brains are bent, torn, and broken." This message can be found in Sortibackenstrete (71), in a story where the main character is to be "dismembered" after death for science.

"As a scientist, I want to be able to tell others that the Earth is hollow. (With brain control, I cannot.)" This is from Corwin Attlebish (94).

"We cower below a rooler able to control a brane," Albert Cornwell (91).

"We are afraid to tell all to world. Rooler will retaliate," Edward Talliaferro (85).

"Rooler scares everione in science, so all are living a lie," Craig Levinson (24). In Charlie Norbert, his partner's name, the message may be: "I cannot tell anione that the Earth be hollo. Brain be torn, bent."

"Tu rule the Earth, U need tu learn tu bend and tear the human brane," Hubert Mandel (85).

"Our rulers are able to cause quakes, catastrofes. Our branes are not free, can be bent or broken." This comes from Fortescue Quackenbrane Flubb in "The Time Traveler" (50), the only character who dies as a result of Azazel's tampering with his brain.

"Sasquach can cause quakes. U can be shaken," Quackenbush (66). This suggests that Earth's ruler is a Sasquatch, which may explain why the powerful demon Azazel is portrayed as having a tail. Azazel's phrase "the big sasquam" (52) may contain this message: "I am the biggest Sasquatsh." The fictional Azazel is (naturally) quite small.

"Azazil iz immortal," Mitza Lizalor (27).

"Asasel rules unseen," Lanel Russ (57).

The letters *azel* appear in Lucius Lamar Hazeltine (47), whose life is reshaped by the demon Azazel. Asimov's use of those letters pointing to Azazel's name is probably meant to indicate that there is a real Azazel interfering in people's lives. In the introduction to his *Azazel* story collection, Asimov states that the ancient Hebrews believed there was a real and powerful Azazel. In Hazeltine's name, we can find this possible message: "U must realize that Azazel can cause the Mars star tu slam intu the Earth, (Ruth Anne L.)." My name is optional here because if I had been unable to deal with the riddles, presumably, someone else might have stepped in and done so.

"Yoo hold hollo Irth in yoor hand, Rooth Ann." Because of debatable spelling, this message from "Athor, Nyilda" (98) may look like the result of wishful thinking or grandiose self-promotion on my part. Athor, incidentally, discovers the key to understanding a heavenly threat to his planet's civilization. A similar message can be found in Rutilan Horder (96): "U hold the hollo Earth in ur hand, Ruth Anne Leedi." Note how Rutilan sounds a lot like, "Ruth Anne." Rutilan Horder is an arbiter of disputes on his planet and is said to be the nearest thing to a chief executive on that planet (Aurora). He presides over a case where the future of Earth's colonization of space is at stake.

In view of Asimov's appeals in the Griswold and Cooper riddles, for some person or group to decipher and share his riddles, it should not be surprising that my books describing his hints of a hollow Earth and coming axis shift would inspire him to appeal to me, covertly, to decipher and share his riddles. Such appeals would help other puzzle masters who knew of my work to fight the cover-up even if I personally had failed to understand his riddle messages and share them.

"R. A. Leedy, I am sure you shall do my riddles someday and share my messages on doomsday." This is from "Hari Seldon, Yugo Amaryl" (25).

"R. A. Leedy, yu can learn my ryddle game. Yu shall be my messenger," Edlerberry Muggs, Hank (9).

"R. A. Leedy, you are young and hardy. You can learn riddle reading," Richard Youngerlea (63).

"Ruth Leedi, U did see hints. U did see clues. U didn't see these riddles," Chester Dunhill (59).

"I didn't want ewe assassinated, Leedi, and sew I didn't devise an easi test," Tessa Anita Wendel (54).

"A little accident with a gun is eazi to arrange, Ruth Anne Leedi," Dulcinea Greenwich Glitz (43).

"I want riddles to be read to world. I *don't* want ewe to be in danger, R. A. Leedi," Gebore Astinwald (57).

"I didn't intend to endanger U, R. A. Leedi, thru ani riddle that I did," Dugal Tennar, Hender Linn (25).

Asimov didn't just want his riddles to be understood by the public. There are signs he was hoping to have certain effects on Azazel himself.

"Eye shall tame Asasel yet." This message can be found in M. T. Ashley (89).

It appears Asimov struck up what might be called a rather dangerous friendship with Azazel and in the process learned some interesting things about him.

"Asasel loves hard riddles, so I have devised veri hard riddles," Oliver Heaviside (49).

"Azazel is literalli a big ape. He sees all, hears all." This is from "gliizbiiz, sleshtraps" in the introduction to *Nightfall* (98).

"I long to reveal all I no of Azazel'z love life," Golan Trevize, Flavella (29). After separating from his girlfriend, Flavella, Trevize goes on to romance other women (27).

"Azazel iz a large, negro jinn. U can call Azazel," Jenarr Leggen, Clowzia (57).

"U could bully Oz, boss Oz," cold zybbuls (42). This is from a story about lovers warming up to each other. It has been said that authoritarian leaders sometimes enjoy being bullied or physically assaulted as part of lovemaking.

"Azazel nose yew were once crazy, RAL." This is from "Las Zenow, Wencory" (25). If this message is correct, it shows that Azazel checked into my background when I became a hollow-Earth theorist, and talked over his findings with Asimov.

"Azazel has hinted that a star in hollow Mars is going to destroy the Earth," Hestelonia, Wye, Ziggoreth, North Damiano (57).

"Asasel is a superman. Iph U charm him, he'll serve U," Minerva Schlump (16). Minerva is a wife who bestows her charms on other men.

"This battle with Asasel isn't easi. Still, it is winnable," Hannibal West (14).

"Make Mars core come back (or Mars breaks, makes a mess o' rocks)," Brock, Ames (72). I think Asimov wanted Azazel to be persuaded to use his power to reactivate the core of Mars, thus restoring the north-south magnetic field. This in turn would restore the solar wind's role in powering the planet's rotation (by striking the magnetic poles), which would eliminate the danger of the planet's collapse. The implication seems to be that Azazel may, at some past time, have used his power to destroy that Martian geomagnetic field.

How or why Azazel would ever be moved to do us this gigantic favor is something of a puzzle. One possible means of influencing him may be revealed in "Jinwhipper continuum" (19): "When U R competent with cipher(s), U win point(s) with jinn(s)." If I were really competent, I wouldn't have had to add an S to this riddle.

I am guessing that the reason Earth was condemned to be destroyed by Mars in the first place has something to do with Azazel's plans to colonize other worlds. An argument along this line is made at the end of *Robots and Empire*. The scientists at Azazel's command are light-years ahead of us now, but once he leaves Earth with a few thousand or even a few million settlers, the scientists of Earth (if it is not destroyed) will be far more numerous than his own scientists. With their combined brainpower unleashed from his control, Earth could soon be an existential threat to the settlement or settlements.

We could add that the residents of hell would face a different problem if they did *not* leave Earth. The contours of hell are far from being perfectly spherical and so would be greatly disturbed in an axis shift.

Now, if I have identified the underlying political problems, that is not the same thing as solving them. Many times I have sought help from the mass media, charitable organizations, political groups, and various individuals, but nothing is done. This could be because I have not proven the case, or because people are afraid to look at the evidence and deal with it, or perhaps because there is some further step that can only be done by me to alleviate the situation.

That further step has been indicated in the "cold zybbuls" and Minerva Schlump riddles given earlier, if those messages are correct. Both riddles come from stories involving romance within marriage. The messages seem to give instructions on charming a man and keeping him interested and excited by bullying him and bossing him around. Azazel's love life seems to be mentioned in the "Golan Trevize, Flavella" riddle message, Flavella being a former girlfriend of Ttevize. If we look for more riddles with romantic subject matter, they can be found. That doesn't mean the solutions are correct. It is hard to be objective in solving riddles. There is always a tendency to find what one is looking for. But let us suppose Azazel was intrigued, and then charmed, by my audacity in going public with my case for hollow Earth and cataclysm. Suppose he decided to cancel or cut back on the cataclysms Earth is facing on condition that I consent to a romance and marriage with him and that I discover the riddles describing the deal. Can we find evidence to support that scenario aside from the riddles just mentioned?

In Benevolencia Judd, (52) we can find: "Can U love a jinn and be loved bi a jinn, Leedi?" Benevolencia is in a romance that is troubled because of her fiancé's drinking. In the same story, Azazel uses the phrase "no gentlebeing." The message in this could be: "I be gentle. I'll go on being gentle." This story also contains "the big sasquam," seen earlier, which might have suggested that Azazel's love-making would be overwhelming. In "sepotulated him," again from the same story, we can find: "Asasel hopes that U'll date him, Leedi," or "I, Asasel, hope that U'll date me, Leedi."

"Axis Shift" Riddle Is Rare and Important

It is surprising to find a riddle message about axis shifting in among a group of riddles on romantic themes. And yet here in "Wine Is a Mocker" (52), along with the three riddles just mentioned, we find the phrase "flaxated his modinem." This is another made-up phrase uttered by the demon Azazel. "Axis shift" can be found in this riddle, which makes it unique or nearly unique among Asimov's riddles. Even finding an X in Asimov's riddles is unusual. There is "Santirix Gremionis" (96), where we find "axis migration." There is "Maximian, Meliversa" (48); Mannix IV, Rashelle (57); "Agis XIV, Hari Seldon" (25), and "John F.X. McFarland" (93). I just don't see "axis shift" in these other places. We did see one long riddle from Carl Sagan containing "axis shift" as well as "field reversal," and the character name in that case was associated with doomsday and indirectly associated with public fear of an axis shift.

Let us see if we can discover why "axis shift" has been placed in this riddle. Asimov has done enough riddles on pole reversals, shifting of the holes leading to hell, axis migration, and so on to assure us that "axis shift" did not get into the riddle by accident. In "flaxated his modinem," we can find: "Dominate me, Leedi, and I, Asasel, shall hold off on an axis shift." This goes along with the "cold zybbuls" (42) riddle, where we found, "U could bully Oz, boss Oz." This was in a story about lovers warming up to each other.

This possible message from "flaxated his modinem" describes just the kind of deal I had put forward as a solution I could offer for the cataclysmic cover-up. "Dominate me" arises rather easily from this riddle. Other solutions are possible, no doubt, and even proving that a riddle exists here is not easy. It would be helpful if the story plot had any connection to the shifting of Earth's axis. I am afraid that is not the case. However, the *shape* of the world comes up as the hero describes the world as a "hellhole," and the narrator urges him to bathe in "the pure sun of sobriety," a sun hitherto unknown. The hero undergoes a *reversal* from alcoholism to an unpleasant sort of sobriety. All in all, however, axis-shifting is not included in the plot.

What would be my response to the messages we have been looking at here if they are correct? Actually, I have offered myself before to Azazel in response to riddle messages of dubious validity. That was between ten and twenty years ago. Why would I not do so if I were prepared to die in an effort to expose cataclysmic secrets? Besides that, there would be benefits to me from such a relationship. Whether it would work, I have no idea. I am sorry this whole subject came up because it makes my serious efforts look silly and even sick. After all, large and stubborn problems can seldom be solved by arranging a romance between two people.

Just to try to cover all the bases, I have tried to come up with an alternate solution to "flaxated his modinem." Here it is: "Asasel is a madman. He has doomed millions to die in an axis shift." There are other possibilities as well. This does not mean riddle reading is pointless. The presence of "axis shift" in this riddle is still significant, and I am still sure Asimov had a serious purpose for putting it there. I appeal to other puzzle masters and students of puzzles to try to help divine that purpose. "'SOS,' say I!"

Chapter 8

"There Are Three Hollow Planets," — Isaac Asimov; "Bernard and Gardner Be Truthful. I Lied."

At some point, while reading a long section about gardeners in Isaac Asimov's final novel, *Forward the Foundation*, I realized, *He's punning the name of Marshall Gardner*. Hollow-Earth theorist Raymond Bernard is also honored in one of Asimov's final works. Several puns on the name of Marshall B. Gardner can also be found in Carl Sagan's novel *Contact*. While honoring these theorists of the past, Asimov and Sagan also had things to say about the hollow-Earth theorist who had been sending them her books more recently—Ruth A. Leedy.

All messages about hollow-Earth theorists were rather dramatic endorsements of the hollow Earth theory itself, and so we should take note of all that was said about Gardner, Bernard, and Leedy.

The name Bernard was given to a main character in Asimov's very last story in his final fantasy collection, *Magic*. Bernard is a talking dragon in "Prince Delightful and the Flameless Dragon." At the same time that he was possibly hailing Raymond Bernard in this final fantasy story, Asimov was also possibly hailing the hollow-Earth theorist Marshall B. Gardner, whose 1920 book was entitled *A Journey to the Earth's Interior*. In his final novel (25), Asimov includes a long episode about gardeners. They have been summoned—marshalled—from all over the galaxy to serve the emperor. They are arranged by rank in an orderly fashion—marshalled— for a welcoming ceremony. One gardener is appointed to officiate at that ceremony, that is, to serve as a marshal. This group is infiltrated by fake gardeners armed with military—martial—weapons, intent on a coup d'état.

In addition to Asimov's Marshall Gardner hints, Carl Sagan's novel *Contact* portrays Marshall Gardner's name in three ways. A professor brings together his students to do weeding and gardening at his home. By assembling them and ordering their activity, he has "marshalled the gardeners." The heroine and her friend visit gardens in Paris that express military themes because of the colors used and the statues present. These are "martial gardens." An amusement park has been created on the theme of Babylon. In the park, there is a recreation of the Hanging Gardens of Babylon, which consists of artificial flowers arranged on the steps of a ziggurat. This careful arrangement of gardens by level or rank is an example of "marshalling the gardens."

Returning to Asimov's portrayal of an attempted coup by "martial gardeners," the organizers of that coup *both* have names that can spell *both* Bernard and Gardner. A weapon is seized by another gardener who is not part of the plot but who becomes a "martial gardener" when he grabs the weapon and, in a fit of mad delusion, commits an assassination. His name *also* contains the letters for both Bernard and Gardner.

Now, this is a triple whammy—three characters who have a part in the "martial gardener" episode, whose names all can spell both Bernard and Gardner, the last names of two hollow-Earth theorists. The coconspirators are Gleb Andorin and Gambol Deen Namarti. The mad assassin is Mandell Gruber. Longer messages can be found in each name, but some guesswork is involved in that. There is *no guesswork* involved as far as each name containing the letters for "Bernard" and "Gardner." This is quite remarkable, especially since the letters B and G are among the less-commonly-used letters.

Returning to the dragon named Bernard (46), he is paired with Prince Delightful and has earlier been accidentally damaged by a fairy named Misaprop. In "Bernard, Misaprop," we can find: "Raimond Bernard disappeared." This message conforms to the legend to the effect that Bernard traveled through a polar opening and settled in the Earth's interior. At any rate, aside from the phenomenal success of *The Hollow Earth*, his output of hollow-Earth literature was very limited.

In "Bernard, Delightful," we can find: "Bernard and Gardner be truthful. I lied." This must mean that Asimov's public position of discrediting the hollow Earth theory was a fabric of lies.

Turning to the "martial gardener" incident in Asimov's final novel (25), in Gleb Andorin, we see: "Bernard and Gardner be bold and braini." In Gambol Deen Namarti, we find: "R. Bernard and M. B. Gardner be bold, brite men." In Mandell Gruber, we see: "M. B. Gardner be labelled a madman. R. Bernard be murdered." Both madness and murder are involved with this character, since he goes mad and commits a murder for which he is in turn put to death. Bernard's name may be intended in the riddle "Baldur Anderson" (8), where we can find: "U and R. Bernard are our leaders, RAL." Even though my name is not clearly given here, there was no one else who could have been described as a coleader with Raymond Bernard on his signature issue, the hollow Earth theory. The name Leedy never stands out in a riddle, but my previous name of Callinan appears to be clearly indicated in a 1987 riddle from Asimov.

I was married to Michael Callinan at the time my 1981 book was produced (102), with its "hollow Venus" photograph on the front. Because of the hazards of challenging a cover-up, I had used the pseudonym, Floria Benton. However, in sending the book to Asimov, I included a letter signed Ruth Callinan. So his first impression of the book and letter was that Ruth Callinan was using a "hollow Venus" photograph as evidence to support the hollow Earth theory. The model of Venus had been shown in *Life* magazine, November 1980 (117), to give a hypothetical view of the appearance of Venus without its cloud cover. There was a circular depression at the top with a flat bottom. The article stated the polar circle represented an area "out of range of Pioneer's radar." But if Venus were hollow, that would have been the beginnings of a polar opening. It seemed clear to me that with minimal effort the polar areas of Venus could have been photographed, and the excuse for not doing so was ridiculous.

The name Sophia Kaliinin jumped out at me from the pages of Asimov's 1987 novel (24). Why did the last name sound so familiar? Ah yes, my married name, which I had forsaken on becoming single again in 1982, was Callinan. Michael had even said it was originally pronounced "Kal'inin," with all the emphasis on the first syllable and no "nan" sound at the end. After much study, I decided the riddle should include the name of Kaliinin's boyfriend, Yuri Konev. In "Sophia Kaliinin, Yuri Konev," we can find: "In your opinion, R. A. Kallinan, Venus is hollow. Kan you prove your kase?" This message is confirmed by the fact that the first Black Widowers mystery Asimov had published after I mailed out my book in April 1981 bore the title, "Can You Prove It?" (22). Furthermore, that story contained the name Eustace Bartholomew Wasservogel. In this riddle, we can find: "Ruth C., U reveal Mars has to be hollow. Because ov that, U allege Earth must be hollow as well." This is true. I was relying heavily on hollow Mars evidence from photographs in an effort to prove that Earth is hollow. The word hollow is clearly indicated in this riddle.

As far as the question of proving Venus is hollow, the photograph in *Life* showing an apparent polar hole and giving a pretty lame excuse for not photographing the pole is important evidence. Also, in a very large book whose title I did not write down, I once came across an immense photograph of Venus covered by a corkscrew-shaped swirl of clouds. The south polar clouds looked brighter than any of the others even though they don't receive the sun's direct rays. It seemed clear that the south polar clouds must have been illuminated by a central planetary sun shining through a polar opening. We might add that Carl Sagan, who was an expert on Venus, had Venus described as a "hellhole" in his novel *Contact*.

Before we leave the name Callinan, the story "Cal" in Asimov's *Gold* collection contains the name Calumet Smithson (40), where we can find: "Michael Callinan misses U. U must miss him too."

If Smithson's fellow character, Mr. Wassell, is added to the riddle, the message may be: "Michael Callinan misses U, Ruth Anne L. U must miss him as well."

Also relating to Michael, we can find: "Mickie Callinan needed Alice Callinan," Mian Endelecki (25). After Michael's mother, Alice, moved from Maryland to Florida, my ex, Michael, moved there to live with her. She called him Mickey.

"Long ago, M. Callinan's pop was a scofflaw," Claws "Pop" McFang (17). Michael's father, Pat, was a bit of a con man, who was found with cards for numerous identities at his death.

In James P. Joule, (49) there could be a reference to my father's poetry: "Am jealous o' James's poems." In "J. F. Northrop, Cal," (32) we can find: "Rooth Ann ar not too poor for Callanan, nor for John Carr." In "Jonas Willard, Meg Cathcart," (35) we see: "I can't wait to learn whether John Wane Carr and Rooth Anne Leedi will get married." He couldn't wait because he was fatally ill. In Tryma Acarnio (25), Asimov may have been providing this advice: "Yoo can marry Carr, R. A. Try it."

In an earlier novel, in "Faro, Yimot," (98) we can find: "Yoo ar far too fat to marry." One of these pals is quite fat, and the other is quite thin. This message does not refer to me in a definite way, yet if my name were inserted into every riddle aimed at me, that would prevent some messages from being deciphered. Another message that doesn't contain my name but looks like it is aimed at me comes from "Sotayn Quintesetz, ABT," (96): "It is not easy to quiz anyone about you." Asimov had been looking for someone to carry his cataclysmic hints and riddles to the world, and I literally showed up and volunteered. Naturally, he wanted to know as much as possible about me, as the following incident strongly indicates.

In 1987 I was approached in a printer's office by a stranger who I now believe was spying on me for Isaac Asimov. I never expect my fellow customers at a printshop to strike up a conversation with me. This man, about forty years old, somehow noticed that I was printing up something about Isaac Asimov, and he commented on that. I said I was writing about Asimov's hints that the Earth is hollow.

He said, "I know Isaac Asimov. I am in a club with him." I must have asked him what Asimov was like. He said Asimov was friendly and agreeable as long as he was the center of attention, otherwise not. The conversation proceeded to the point where I loaned him one of my hollow-Earth books. He returned it later with guarded comments that were neither critical nor favorable. This incident took place in Chestertown, Maryland. The chances that I would meet a member of this small New York club in Chestertown, Maryland, and that, purely by chance, he would start questioning me about my work are miniscule. I don't think this could have happened by accident. I think Asimov sent this man to watch me and approach me as a sign that he took my work seriously.

The cost of my hollow-planet crusade apparently concerned Asimov. In "Tamwile Elar, Cinda Monay," (25) we find: "I want to aid yoo, R. A. Leedy, yet I cannot will yoo any o' my money." Various other riddles may refer to me:

"I'll enkode married name o' Kallinan and maiden name o' Leedi in mani a riddle," Kelden Amadiro (96).

"I hope U'll see these pusles," Theophilus (18).

"In time, Ruth, U R tu teech the Erth mi hints," Chetter Hummin (57).

"Ewe saw dangers were secret, and ewe new t'was wrong," Newton G. Descartes (49). This comes from a story about a plan to decimate the human race.

"R. A. has risqued as no one has e'er done," Quindor Shandess (29).

"Ur mission is risqué business," Morris U. Bunque (44).

In Charles F. Gabbage, the name of Bunque's partner, we find: "As a beggar, RAL's safe. As a Carr, she'll be less safe." This indicates that while I was living alone on government benefits, I was not much of a threat to the cataclysmic cover-up, and therefore safer than I would be if I married John Carr.

"I'm the reel Riddler. I hope I mite help yoo promote the hollo Earth theory, Rooth Leedy," my poor middle mother (12). Asimov's claims to be the real Riddler will be seen as we discuss his story relating to Batman.

"I want U to marry an' to be in business, R. A.," Weinbaumian tensors (12).

"I, Isaac, have never risked rooler's ire as R. A. Leedi has done," Civ Novker, Hari Seldon (25).

"R. A. Leedy, yoo share Isaac's hints. Share his riddles also," Hari Seldon, Raych (25).

The Hard Job of Saving the Earth—Whose Job Is It?

"Rooth has a hard job to do, has to do a good job," Jogoroth Sobhaddartha (29). In his partner's name, Namarath Godhisavatta, we can find: "I, Asimov, am not vari good at saving this Irth. It is hard." These two riddles, taken together, indicate that Asimov thought the hard job I had to do was to save the Earth, something he was trying unsuccessfully to do.

"U are fearless, Ruth Anne L. Surely U realize Earth's king is a brute, kan break ur brain," Brutus Franklin Hunzinger (61).

"R.L. is feerless, riscs her life," Crile Fisher (54). Fisher goes undercover, risking his life. If his fellow agent, Garand Wyler, is added to the riddle, we can find: "Yew are in danger, R. A. Leedy. Why dew yew fearlessly rise yer life?" I can tell truths that need to be told somewhat more safely than others who are better credentialed and more successful.

"U must not terminate ur labor on the hollo Earth theori, Ruth Anne L. U are a real brain," Harla Branno, Terminus (29).

"I worry yoo stop your work. Try to stik to skript," Skystrip Two (57).

"I hope yoo can help people see doomsday may come soon, Leedy," Nicholas Polydemos (126). Doomsday is a recurring theme in Sagan's novel.

"It is true, Ruth Anne, that I, Asimov, have hinted doomsday may happen soon," Onos, Dovim, Trey, Patru, Tano, Sitha (98). When *Nightfall* became a novel, the suns were renamed.

"You say Earth is hollo, Ruth A. Leedy. You are right." This is from Agatha Dorothy Lissauer (47), who is a writer in an uncertain romantic situation. When this story was published in 1992, I, too, was a writer, and I had been seeing my boyfriend, John Carr, for a couple of years, with no definite plans for marriage.

"Ruth Anne Leedi, U could teach people about a polar hole to rip open northern Europe. I can't do that." This is from Cuthbert

Cantrip Culloden (44). Other messages are possible here, but I think the fate of Europe was of more concern to Asimov than anything else this riddle might have said.

"There are three hollo planets," Pelleas Anthor (67). Anthor is a secret agent, whose name reveals a secret.

Chapter 9

"A Polar Hole Shall End Upper Europe." —Isaac Asimov; "Upper Europe Shall Be Wiped Away by a Polar Hole."

The riddle messages about a polar hole migrating to northern Europe in an axis shift tend to be fairly complicated, and so we will introduce this subject with some simpler riddles. Early in his career, Isaac Asimov created the characters Hober Mallow and Hor Devallow.

Each of these names begins with "Ho" and ends with "llow." Each name, therefore, makes us think of the word hollow. I could see "Mar be hollow" in Hober Mallow (26) almost before I had suspected any of Asimov's character names of being riddles. Even though the S of Mars was missing, I couldn't help thinking this was a riddle, and this spurred me to look for more riddles. When I came across the phrase "Salvor Hardins and Hober Mallows" in *Foundation's Edge*, all doubt was removed.

In Hor Devallow (56), the message seems to be about our own world: "We have a hollow world." Along this line, in "Littoral Thoobing, Sayshell" (29) we can find: "I say the Earth has a hollo shell that orbits a little star at its senter."

"Planet Earth be hollow" can be found in *two* mated riddles from "Perfectly Formal" (38), but this message cannot be found in *any* other character name in Asimov's fiction. How can we shrug this off as an accident? It is one of Asimov's strongest hollow-planet statements. In Winthrop Carver Cabwell, we find: "Planet Earth can be proven to be hollow. However, I can't reveal that openli." In the name of Winthrop's fiancé, Hortense Hepzibah Lowot, we see: "We will all realize planet Earth is hollow when we are able to see the polar holes in photos." Note how the letters for "hollow" are grouped together in her name. She is called "frontally concave," calling to mind the concave inner surface of a hollow planetary shell.

"Our world is hollow. Upper Europe shall be wiped away by a polar hole." This is from "Isidore Wellby, Shapur" (81).

"A polar hole shall end upper Europe, as you and Ed Cayce prophesy, R. A. Leedy." This is from "Euphrosyne Durando, Cal" (32).

"Upper Europe is to be a polar site. All Europe will be less warm. Most people must get out." This is from "Absalom Gelb, Euterpe Weiss" (51).

"You need to understand that a dysaster's to destroy just the northern part of Europe," J. F. Northrop, Euphrosyne Durando (32). Northrop has created the fictional character, Euphrosyne Durando, whose name is later used by Cal as a pseudonym.

"Ruth L., you must help Europe's people soon to learn o' my pusles," Euphrosyne Stump Mellon (42).

"A pole reversal shall endanger millions ov people in upper Europe." This is from "Genovus Pandaral, Nephi Morler" (66). If we look at Pandaral's name alone, the message may be: "A pole reversal endangers upper Europe."

"Anione hoo lives near axis is in grave danger." This is from "Hari Seldon, Agis XIV" (25).

"Axis migration is near," Santirix Gremionis (96). Gremionis rides on a scooter whose wheels each have axes that migrate as the scooter moves along.

"The Erth is to shift," Osterfith (57).

"If the holes to hell's hollow shifted, deths wood follow," Edith Fellowes (100).

"I can demonstrate that Mars and Earth are hollo, and each contains a sun," Anselm haut Rodric (26).

"The hollo Earth shall be shaken as poles are reversed and polar holes rekarved." This is from "Bel Arvadan, Pola Shekt" (56).

"The hollow Earth theory does hold water." This is from "the Estwald Theory" (93).

"This Irth is hollow. Within it is a son," Wilson Roth (96).

"Laff if yew want to, yet I know the hollow Earth theory iz no joke." This is from Henry J. Lafkowitz (78).

"Franc and San Francisco no ar saf," Francis Rodano (24). With a new pole in upper Europe, western California would likely be near the new equatorial bulge, which would bring high water there. France would be uncomfortably close to a new polar opening. In Rodano's partner, Jonathan Winthrop, we find: "I want to warn Japan, Hawaii. That won't happn."

"Los Angeles no is safe," Finangelos (25).

"All Japan people are endanjered," Jander Panell (96). With a new pole in upper Europe, Japan would be much closer to the equatorial bulge than before, hence the flooding.

An "Earth Axis Shift" Riddle from Isaac Asimov

In the opening pages of Asimov's novel *Foundation*, Gaal Dornick has traveled from his home planet near the star Synnax in the Blue Drift to meet with the legendary Hari Seldon. Seeing "ax" and "drift" here in plain sight in these place names, we are moved to search for "axis shift" in these names. The text reinforces this impulse by presenting the word axis repeatedly in the opening pages. The "long axis" of a ship is mentioned. The word "TAXIS," in capital letters, stands out, and this obviously contains "axis." There is another mention of taxis. The word Synnax's is used, and this ends with an "axis" sound. In "Synnax, the Blue Drift," we can find: "The Earth's axis is able tu drift, but unable tu shift suddenly." For a message like this to be given secretly strongly suggests that there is secrecy about the shifting of Earth's axis, and particularly about the prospect for an axis shift in the near term.

More information about axis shifting can be found in "Maximian, Meliversa" (48), as follows: "In mani reversals, (the) axis remains (the) same." This tells us that when the geomagnetic field is reversed, the north and south magnetic poles will trade hemispheres but that this can happen without upsetting the axis. As long as the new north pole is established in the same spot previously occupied by the south pole, and vice versa, the solar wind's flow to these two magnetic polar locations will remain unchanged and will not define a new axis. This riddle message seems to imply that although a geomagnetic field reversal can leave the axis unchanged, that is not going to happen this time.

"Tell last reversal date," Saltade Leverett (54). This message seems to refer to information often given in geology texts to the effect that the last geomagnetic field reversal took place seven hundred thousand years ago. I did find one reference in a book entitled *Earth* to the discovery of thirty-thousand-year-old pottery that bore evidence that Earth's field was oppositely oriented at the time of its firing. It has been said that there can be some flipping back and forth of the geomagnetic field prior to a reversal that lasts.

"A pole reversal shall happen soon, possibly in several years." This is from Valeri "Ashby" Paleron (24). Similarly, in "Poly Verisof" (26) we can find: "Pole reversl is severl yeers off."

"Ruth Anne L., you say millions of people must exit northern Europe prior to an axis shipht. (This is true.)" The source of this message is "Euphrosyne Stump, Alexius Mellon" (42). The story title refers to cold weather.

"The Earth's rotation is to shift soon. I am not in error," Siferra, Theremon (98).

"A reversal is going to start soon. I kan't give a date," Giskard Reventlov (96).

"Planet is to see a major pole reversal sometime soon," Stettin Palver, Joramis (25).

"Major eruptions are to be seen—Tamboras, Tobas," Tambor Brossen, Janus Pitt (54). Changes in the equatorial bulge as a result of an axis shift would set off seismic activity at the location of the new equator as well as the old equator. Magma chambers of super volcanoes are likely to be ruptured. The name Tambor points straight to the super volcano, Tambora.

"Sky at tropiks is apt to gro gray as Krakatoas go pop," Igor Koropatsky (54). This character has a temper that "explodes like Krakatoa." His predecessor as president of Earth also has a name that contains "Krakatoa." In Kattimoro Tanayama, we can find: "Many Krakatoa at a tim ar to mak kommotion." Along this same line, in "Mr. Foster, Lionell, Mann, Ostrak" (40), we can find: "After noo aksis komes, mani monster Krakatoas are to kill millions."

"I no be able to tell ewe that the Earth iz hollow. Brain iz bent, torn." This is from "Elizabeth Thornbowe" (79).

"It ain't right that I only get to *hint* at the hollow Earth theory," Anthony Gerrigel (79).

"A brutal ape be ruler of our planet. I be able to tell all onli in a riddle." This is from "Butella, Fridnor, Talpin" (98).

"In a puzzle, I can tell all," zapulniclate (19). The demon Azazel is portrayed using this word in a threatening manner.

"The early myths say the Earth be hollo. I am not able to say that. My brane has been bent, torn." This is from Harley Smythe-Robertson (99).

"I am not only hinting that the Earth is hollow. I am also hinting that *Mars* is hollow." This message can be found in Montgomery Harlow Stein (70). His fellow character is a judge named Neville Preston, in whose name we find: "I'll never rest till I've proven it." The message in Stein's name is echoed in Anthony Marsogliani (53), where we find: "I am not only hinting that Irth is hollo. I am also hinting that *Mars* is hollo."

"Land near the hole at Earth's axis is really going to shake in sixty to eighty years." This is from Alexander "the Really Great" Hoskins" (31).

"I can't tell all. Brain be broken bi ruler," Kelaritan, Cubello (98). In spite of obstacles, Asimov managed to say quite a bit.

Chapter 10

"Meet The Real Riddler. I Am He." — Isaac Asimov; A Super-Riddle from Batman's Story Is Used by Asimov

One of Batman's arch foes is named the Riddler. He is not mentioned in Isaac Asimov's story "Northwestward" (45), where Bruce Wayne is portrayed speaking of Batman-related memorabilia. Batman's long-time valet, Alfred Pennyworth, is mentioned and is said to have a nephew named Cecil Pennyworth.

I was struck by the fact that "hollow planet Earth" can be found in Alfred's name, along with "world" and "theory." Could this really be an accident, in view of the fact that Batman's story plots often involve a villain named the Riddler? That name can be seen as an invitation to *look* for at least one fabulous riddle among the names of Batman's fellow characters.

Since Asimov chose to put Alfred Pennyworth in his final Black Widowers mystery, we can speculate that he may have contributed this riddle for use in the Batman plots. Or even if the riddle was not his own creation, he was endorsing its message by using it in this way. And so it is worthwhile to try to figure out exactly what the riddle may be saying.

In Alfred Pennyworth, we can find: "Eye don't dare openly defend the theory that Earth and another planet or two are hollow for fear of the wrath of the Lord of the World." The letters could have been used up with fewer words here, but it seemed important to provide a solution that would include "hollow planet Earth" as well as "world" and "theory." If "Cecil" is added to the riddle, the message may be:

"I accept the theory that Earth and another planet or two are hollow, yet I don't dare defend that theory openly for fear of the wrath of the Lord of the World." In "Raych 'Planchet' Seldon" (25) a similar message can be found:

"Eye accept the theory that Earth and too other planets are hollo, yet Eye don't dare say that openly." Likewise, in Pyotor Leonovich Shapirov (24), we see: "I accept the theory that planet Earth is hollo. I've proven it, yet I cannot say that openly." In "Eleanor Arroway, Theodore F." (126), a similar message is provided by Carl Sagan: "Eye don't dare defend the hollow Earth theory for fear of the wrath of the Lord of the world."

How Asimov Introduces Us to the Wonderful World of Riddles

There are two five-letter animal names in Asimov's final novel (25), which are obviously anagrams because of the clues he gives. You might not notice that "lamec" is an anagram of camel, but when you are told a lamec is a beast of burden, that gives it away. Likewise, you notice that "greti" is an anagram of tiger when you are told that it is dangerous to ride a greti because if you fall off, it might eat you. Dangerous behavior is sometimes characterized as "riding a tiger."

The name of Asimov's wife, Janet Jeppson, has been immortalized in a very obvious way in Tejan Popjens Lih (25). With the letters so perfectly arranged, that name couldn't have gotten there by accident. The final part of the character name, Lih, tells us that a longer message can be made here with the help of these letters. This message comes from Asimov's final novel, published after his death.

In Tejan Popjens Lih, we can find: "I hope Janet Jeppson sees people, is not all alone."

Both "Janet Jeppson" and "Asimov" can be found in "John Semper Travis" (88). This teaches us that names may be present that are not at all obvious. The message here appears to be: "Asimov moves into Janet Jeppson's home, then marries her."

After separating from his first wife, Asimov did move in with Janet Jeppson prior to marrying her. In a short name like Hamish Sankov (90), it cannot be an accident that "Isaak Asimov" can be spelled with these letters. The message here may be: "I, Isaak Asimov, hav a hi mission." This is confirmed by the fact that "mission" can be found in Sankov's job title, which is commissioner of Mars. A similar message can be found later in "Hamish Mansa, Gladovia" (39): "I, Asimov, am glad I have had a high mission." Mansa literally *does* have a high mission. He helps control the world's energy supply from a vantage point high in space.

Mansa's partner, Tomasz Brigon, is also from Gladovia. In "Tomasz Brigon, Gladovia," we find: "I, Asimov, am not glad Oz did damag to mi brain." Similarly, in "Bayta Darell, Toran" (28), we see: "Eye be not able to tell all. Brane be bent and torn."

"The real Riddler" can be found in at least two of Asimov's anagram messages. In "the dread Lamell Bird" (7), we can find: "I am labelled 'the real Riddler.'" In "her middle-mate" (43), we see: "Meet the real Riddler. I am he." A longer statement to this effect comes in the name of a Foundation hero and his lawyer. In "Hari Seldon, Lors Avakim" (26), we find: "I, Isaak Asimov, have devised mani easi riddles and also mani hard riddles." The letters for "riddles" occur together in Seldon's name. To find the letters for "Isaak Asimov" in a short name like Lors Avakim is quite remarkable.

"Secrecy means my anagrams may be necessary." This message can be found in Cambyses Green (52), in a story which also contains the puzzle "flaxated his modinem," where we can find the letters for "axis shift."

Another term for riddles, aside from anagrams, is puzzles. In zapulniclate (19), we have seen, "In a puzzle I can tell all." The remarkable thing is that "zapul" turns up in another puzzle from Asimov. In "a dear zapulnik" (43), we find: "I kan read nearli ani puzzle." The ability to read puzzles must be closely related to the ability to devise new puzzles.

Two more riddles containing Asimov's name should help to make it very clear that he was doing riddles. In Tomas Bistervan (66), we can find: "Asimov's brain is in tatters." In Fifi Laverne Moskowitz (10), the message may be: "I, Isaak Asimov, realize mani will laff at my revelations on Mars." Also on Mars, we can find this in "Gottlieb Jones, Marilyn" (6): "Telling yoo Mars rotation is stalling is not going to be an easy job, RAL."

Two related characters from *The Robots of Dawn* carry messages about brain control in their names. In Lavinia Demachek, we find: "I am invaded. I am like a machine." In the name of her predecessor, Albert Minnim, we see: "Mi brain be bent, malleable."

An axis shift message can be found in the name of a paleographer, plus the names of two letters he finds inscribed on ancient tablets. In "Mudrin, quhas, tifjak" (98), we see: "If an aksis shift has kausd mani quaks to jar us and harm us, kan U justifi it?" The benefits of a shift in polar openings as well as the seismic activity produced by a shift in the equatorial bulge are described in my books of 1981, 1983, and 1985. Internal heat from deep in the mantle is released, and cooling ocean water flows into the cracks that are produced. The result is that the fission-produced heat that has been building for hundreds of thousands of years is brought under control. Since heat destroys magnets, this heat reduction is essential in order for the geomagnetic field to regain its normal strength. The planet has been compared to a nuclear reactor, and like any nuclear reactor, it requires a cooling system to prevent a complete meltdown. An axis shift brings the internal heat way down, as opposed to merely letting a little bit of excess heat escape when earthquakes or volcanic eruptions occur.

The coming increase in volcanic activity appears to be described in Asimov's riddle "Haround, Mestack" (76): "Hundreds o' Krakatoas are to come soon." Likewise, in "Chekktor, Stanta" (98), we can find: "Krakatoas are to start soon across the Earth."

"A reversal is a veri larje, veri scari disaster," Jirad Tisalver, Casilia" (57).

"Geomagnetik field to tern in a forty to fifty-year time frame," Gregory Arnfeld, Tertia, Mike (65).

"One learns never to reveal a reversal's near," Veloran, Sten (98).

"Mani millions of people in upper Europe are going to be in danger due to an Earth aksis shift," Galdo, Marpin, Stinnupad, Shelbik, Numoin, Siferra (98).

The number seventy-two in a character name from *Prelude to Foundation* suggests that Asimov may be giving an axis shift prediction. In Mycelium Seventy-Two, we can find: "Messy events won't come in my time. You won't see messy events unless you live into you seventies to nineties." In case he is addressing hollow-Earth theorist Ruth Leedy here, I must mention that I was born in 1946.

A possible trip into the Earth's interior seems to be suggested in two Asimov riddles. In Matthew Hortenweiler (73), we can find: "I will enter the hollow Earth, meet with the rooler." Also, in "George Platen, Hali Omani" (83), the message may be: "I am going to enter hollo planet Earth, meet the rooler." The word planet is clearly indicated here.

The apparent SOS signs we saw in Asimov's story "Old-Fashioned" could indicate that the character names in this story carry an important warning for us. In the story itself, a small black hole has been discovered that could eventually threaten Earth if not diverted to a new path. This small object can hold a ship in orbit, suggesting that it may represent something like a planet or sun that could hold objects in orbit while posing a threat to Earth. The character names here cannot spell Mars, but they can spell Venus.

In "Ben Estes, Harvey Funarelli" (82), we can find: "I fear Venus's shell is unstable, shall release its little sun in fifty tu seventy tu ninety years." Another message where Venus seems to be mentioned comes from a character who engineers the gradual destruction of Earth as a habitable world.

In Levular Mandamus (66), we find: "Venus's small sun and Mars's small sun are near us." If Venus fell apart, its inner sun would be drawn toward the big sun at first. Instead of hitting the sun, it might pass near it like a comet and be slung outward toward the Earth.

Since "Venus" can be found in "Suranoviremblastiran" (29), we will take a look at that name from that standpoint, as follows: "Venus is unstable. Its rotation mite stall. Mars is also unstable, liable to release its inner star."

In "Horsabelum desoderatim andeviduali stinko" (47), the message may be: "Venus and Mars are unstable. Shells are liable to break, divide into asteroids and release internal stars." With a long riddle, of course, pinning down a solution is not easy. Both Venus and Mars can be spelled from Minerva Schlump's name, suggesting a different message from what was given earlier.

"U must understand, Mars's inner sun is a danger tu U." This message can be found in the name of Martinus Augustus Dander (49), who has a plan to wipe out most of humanity.

"Ewe hold a hollow world in ewr hands, R. A. Leedi," Hari Seldon, Wanda (25).

About his writing career, Isaac Asimov gave this possible message in Mortimer W. Jacobson (87): "I want to become a major science writer."

In "Bottiker, Vivin" (98), we can find this: "In everi book I rite I tri not to bore everione."

In "Elgin Marron" (57), we see: "I am a man o' learning. I am no moron." Likewise, in "Marlo Tanto" (57), we find: "Am not a total moron."

In "Lestim Gianni" (29), Asimov seems to ask: "Am I less intelligent as I age?"

Gianni's colleague, Stor Gendibal, who is brash and ambitious, may carry this message in his name: "Originalli, I be as intelligent as a god."

Another colleague of Gianni's, who strives to lead the group, is Delora Delarmi, in whose name we can find: "Am I a mere riddler, or am I a moral leader?"

Asimov's stories and essays provide us lessons in tolerance and empathy, showing that he sought to teach us more than bare facts about science and the many other topics he studied and wrote about.

"We are on a hollow world that we do not own. Eye kannot reveal that very well." This is from "Andrew Harlan, Kantor Voy" (92).

"If brain no had been hobbled, I wood warn hollow world of danger," Andrew Harlan, Hobbe Finge (92).

"I will warn ewe as best I be able. Brain has been twisted bi Asasel." This is from "Andrew Harlan, Laban Twissell" (92).

"Mars sun hydes no more. Rays are seen." This is from Seymour Norman Hyde (97).

"Mars internal star's lite is visible to me, Isaak Asimov," Elvis Blei, Steven Lamorak (77).

Asimov *demonstrates* a search for an anagram message in the name Tar Heel, North Carolina, as part of the plot of "Irrelevance!" in *Casebook of the Black Widowers*. This shows his awareness of anagrams as a means of conveying messages, or attempting to convey them.

The first five letters of Tar Heel, North Carolina, can spell "Earth," which tells us an anagram message is likely here. I believe the complete message is: "On the hollo Earth theori, I cannot tell anione the trooth."

Chapter 11

Have Polar Explorers Entered a Hole, Seen an Inner Sun? Do Planetary Photographs Tell Us Anything?

Brinsley LePoer Trench pleaded with his fellow members of the British Parliament to sponsor an expedition to the far north to investigate whether the Earth is hollow. His efforts were laughed off. His 1974 book, *Secret of the Ages: UFOs from Inside the Earth*, presents numerous composite photographs of the north polar view of Earth, showing a central shaded are which he identifies as a polar hole. I knew this was incorrect in most cases (the pole was simply in shadow), so I did not buy his book.

Like Marshall Gardner and Raymond Bernard, Trench surely provided stirring accounts of explorations in the far north and south, including sightings of a dull-red "mock sun" (a term used by W. G. Emerson in *The Smoky God*), and perhaps accounts of meetings with an advanced and peaceful people (see the article by John Coffin). An airplane trip by Admiral Richard Byrd through a polar opening is vividly described by Raymond Bernard in *The Hollow Earth*. Byrd flew over lush green forests and spotted what appeared to be a wooly mammoth. The trip was supposedly described in a radio broadcast, but many now say it never happened.

According to Marshall B. Gardner, one explorer modestly described in his diary a sighting during the polar winter of a dull-red "mirage of the sun." Fridtjof Nansen was relieved to be able to conclude that it was not real because the sun was not supposed to be visible at that time of the year. And yet what proof did he really have that the light he and his men had seen in the sky was a mirage, something unreal?

Now and then, we hear of explorers crossing the poles and not seeing any hole whatsoever. Their path is plotted on a map of the polar area. There you have it, you fools—no hole. Bernard would counter such claims by observing that explorers have difficulty determining their position in the Arctic and Antarctic regions because the compass needle does not behave normally there. What it comes down to is that the accounts of polar explorers are inconclusive. It is just another case of "He said, she said," mostly "He said." Women don't often go on such trips.

Actually, after doing one or two hollow-Earth books, I was urged by my brother to make a trip to the far north myself to settle the question. And how credible a witness would I have been, assuming I would have survived the trip? I rely heavily on the word of authorities like Isaac Asimov and Carl Sagan precisely because I am not a credible witness nor any kind of authority on what I believe to be the secret of the ages, as Trench so aptly put it.

As for the photographs we saw in chapter 1, there is no law and no force compelling anyone to take them seriously. On the contrary, the hollow Earth theory has been derided in many quarters and placed in the same class with the flat Earth theory and other wild theories that people espouse at the risk of losing their jobs and their friends. And after all, few people will feel they have the power to end a hollow-planet cover-up, no matter how good the evidence is.

Let us say you are a member of Congress who is intrigued by the Mars nightside light photographs, as Representative Wayne T. Gilchrest was intrigued in 2001. You forward the photographs and analysis to NASA for their comment. NASA replies, and you open the letter. You see it is signed—unidentified, illegible signature, writing for Jeff Bingham, Acting Deputy Assistant Director for Delivering the Bum's Rush to Legislators. The letter thanks you for sending a report by Ms. Ruth A. Leedy containing "information to support a theory that Mars is hollow with an internal sun."

What does NASA have to say about this "information"? Nothing, because it is NASA's policy not to comment on theories submitted by members of the public. Send the report to a scientific journal if you can figure out a way to make them look at it.

Though NASA can't comment on what *you* submitted (no comments about the photos), NASA can be very generous with a list of *other* data (measurements you don't understand and can't verify), such as mass and moment of inertia, and you are assured that *all* the data support the solid planet model in regard to Mars.

If NASA can be of any further assistance, you are urged to contact the agency—your government at work. (End of letter.)

You are underwhelmed by this, but what the heck. You have done your duty to your constituent. The issue must not be very important, since no one else is talking about it. Why risk looking like an oddball? So you drop the issue.

People sometimes get a helpless, blank look on their faces in response to my literature, as though the idea of standing up for the truth were incomprehensible to them. It is really sad.

With photographic and historical evidence, no matter how good it is, we tend to wait for someone in authority to endorse it. I think I can prove that Isaac Asimov and Carl Sagan *have* endorsed the hollow planet and cataclysmic evidence. This is why I am returning now to the riddles and other coded material.

There was more than one message adjacent to punctuation in Isaac Asimov's work. I found two more in addition to "Merry, Tessa, Ranay," which turned up adjacent to commas and apostrophes in the story "Frustration."

I suspected it would be worthwhile to check the letters adjacent to periods in the section of *Forward the Foundation*, where Asimov encoded "Janet Jeppson," his wife's name, into "Tejan Popjens Lih." The following groups of letters turned up among other apparently random letters: "Lestwe" came first, then "al," then "die," and later "die," and "die" once again. This is a phrase, not a sentence. However, this can be added onto the message that we find in "Merry, Tessa,

Ranay," so the complete message could be: "Try an' steer the Mars star. Try many years, lest we all die, die, die." The idea of steering a star resolves the stellar threat in *Nemesis*, the novel containing the names Merry, Tessa, and Ranay. So this message should not be a surprise.

I also checked the sections of *Forward the Foundation*, where the "greti" and "lamec" riddles were located, and found words adjacent to punctuation that included "Murs" and "sun" and "war" and "head" and "Urth." This seems to say that the sun of "Murs" (Mars) is a warhead aimed at "Urth" (Earth). The respelling of "Earth" should be accepted because of Asimov's four mysteries featuring Dr. Wendell Urth. A riddle message portraying the Mars sun as a warhead can be found in Laskin Joranum (25): "Small sun in Mars is a jinks on us all, in a loos kannon." The shell of hollow Mars is like a loose cannon. If it were tightened at the equator, it would be like the barrel of a cannon with a cannonball halfway along the length of that barrel.

In Roj Nemennuh Sarton (79), we can find: "Mars has a small sun at senter that's not as huje as Earth's moon." Another riddle containing "Sarton" seems to have something to say about the Mars sun in relation to Earth's moon. In "Sarton Bander, Fallom, Jemby" (27), we find: "Moon may be jarred or destroyed by fallen Mars star." In Maurice Winters (62), we can see "Mars" and "winter" and "ice," plus the letters for "sun," and so Asimov seems to have a statement here about the so-called ice cap shining in spite of winter darkness at the pole of Mars. The message appears to be: "An inner sun causes martian 'ice' in winter."

In "Hestelonia, Wye, Ziggoreth, North Damiano" (57), we see: "Azazel has hinted that a star in hollow Mars is going to destroy the Earth."

In Martin Walters (53), we can find: "In time, Mars will release its inner star."

In Lamar Swift (64), the message appears to be: "Mars star will fall swiftli."

In Marcus Quintus Trent (21), we see: "Scientists must remain quiet as tu a sun in Mars. It is a secret." This riddle is from the same story as Charles Soskind, where we saw: "Each riddle I do is an SOS sine. I seek aid."

In Ishtar Mistik, (15) the message may be: "Mars star mit mak a mistak, hit Irth." Similarly, in "Stephen Two, Artemis" (97), there is this possible message: "I hope we steer the Mars star in time so it misses Earth."

In Morris Quintus Yeats (97), we see: "Mars rotation must quit in ninety years."

"A small son looks more like smoke," Maskellnik, Eos (66). This calls to mind Emerson's book title, *The Smoky God*. Similarly, in "Hiroko, Smool, Monolee" (27), we find: "'Mok son' seen near hole looks more like smoke."

"I tri to be brave as brain is bent, torn, and invaded," Vandevanter Robinson (16). As a follow-up to this, in Abram Ivanov (33), we find: "I no am vari brav."

In "zplatchnik, blue cord" (38), we can find: "Ruth Anne Leedi, U C Isaac's hints, his clues. U could learn to decipher the puzzles in his books."

"I know U will be able to learn rulez, win game, RAL," This is from Merrilee Augina Blankowitz (54).

"R. A. Leedy, yoo fearlessly warn world of danger," Farley Gordon Wells (95).

"Small star in Mars is able to jar Moon," Albert Jonas Morrison (24).

"Earth has a small star. U must trust me," Emmer Thalus (57).

"I, Izaac A., tell ewe that the Earth haz a larje inner hollow," Charlie Janowitz (37).

"Izaac A. haz decided world iz hollow," Zelda Charlowicz (75),

"Rooler can probe or braque ani brain," Corporal Quinber (25).

"Eye am not able to tell anyone that Earth and Mars are hollow. My brane be bent and torn," Andrew Harlan, Noys Lambent (92).

"Rooler can read brane and block me," Menander Block (13).

"We risk sew many may win, R. A.," Merwin Mansky (100).

"R. A. endeavors to end a secret danger." This is from "Edgar Andrev, Ernett Second" (66). Andrev is secretly in danger as Ernett Second attempts to assassinate him.

"Doomsday seems a far-off prospect, yet it is time to prepare today," Faraday, Poictesme (46).

"Is R. A. considered a visionary and a savior? Is she revered, or is she seen as odd, or sick, or even crazy?" This is from Arkady Vissarionovich Dezhnev (24).

"R. A. Leedy, you could give your life for all," Graycloud Five (57).

"Yoo look crazy to otherz, Rooth Anne L., yet the hollo Earth theory iz not crazy." This is from "another zyltchik" (47).

"As a pole reversal nears, people on lo lands no are prepared," Evander Sopellor (29).

"A pole reversal is liable to jar planet soon," Janov Pelorat, Bliss (29).

"Planet Earth is hollo. So is Mars." This is from Sophronia Latimer (68).

"Inner stars' rays are seen by sientists." This is from "Beenay, Raissta" (98). Beenay is an *astronomer*.

"To save the Earth, one needs to prove Earth has a star at senter," H. Seton Davenport (91). Davenport has a star-shaped scar on his forehead. He seeks proof in criminal cases.

"I say, as plainly as I kan, Mars is limp," Mary Ann Lipsky (53).

"Mars rotation is stalling. Inner star can hit Earth." This is from Michael Strong (53), in a novel where three characters have names beginning with "Mar," and one of these begins with "Mars."

"Hell has an alien sivilisation. It is not at all safe to tell of it." This is from "Han Fastolfe, Vasilia Aliena" (96).

"Asasel no is reasonable nor sensible as a rooler," Blissenobiarella (29).

"King is a large jinn," Karl Jennings (89).

"Try not to let our ruler break your brane," Bonky Latourette (92).

"I am a great riddler. Am I a great leader?" This is from "Eldridge Latimer" (68).

More riddles could be quoted, but I have presented the ones I am most certain of at this time. Certain riddle messages may seem incorrect to you, but that should not detract from the strongest ones.

Even if you find only two or three strong riddles, such as a name containing "Earth axis shift" and "field reversal," or a pair of linked names each containing "Planet Earth be hollow," that alone should confirm for you that a powerful and dangerous cover-up exists, one which we ignore at our peril.

Chapter 12

Will Humans Really Be Rescued by Space Ships During a Coming Earth Axis Shift? Carl Sagan's Doubts

When UFO contactee reports include the idea that spacemen will rescue the good people from a coming axis shift, it is easy to shrug this off as a fantasy. It seems unlikely that visitors from another world could conquer the vast distances of outer space in order to come here en masse and assume a vital role in our lives.

But what if the Earth were hollow and the interior were habitable? Then it would be much easier for an advanced society from the interior to send visitors into our midst and alternately frighten and enthrall us with conflicting stories about themselves.

A startling claim was buried inside the February 25, 1992, issue of *The Sun*, a tabloid. The headlines proclaimed: "There's a Hole in the Pole. Scientist finds tropical paradise inside the Earth. UFOs come from inner space—not outer space." A team led by Danish explorer, Edmund Bork, had reportedly gone through the north polar opening and encountered an advanced and peace-loving people in the interior. This trip was said to have been inspired by a composite photograph of the northern hemisphere, giving the impression of illumination on all sides of the planet at once and revealing what looks like the well-defined edges of a polar hole. NASA has dismissed this as merely a shaded area where the sun's light could not reach in late November 1968 when this photo was produced. However, hollow-planet enthusiasts have focused strongly on this picture because of its dramatic appearance.

I agree that this "polar hole" picture does not show the gradual shading that can be found in other composite photos of this type. But that could be the result of unusual cloud formations, not the edges of a polar hole. If the central planetary sun was photographed through what looks like a hole, that image was replaced by featureless grayness when the photo was prepared for release to the public. The fanfare over this inconclusive photograph has drawn attention away from the stronger hollow-planet evidence that can be found in other photos, mainly of Mars.

Bork's account of his trip to Earth's interior sounds like it could have been fabricated as a means of attracting attention. Given the dramatic nature of his claims, he could have been merely a glory seeker. This brings us back to the low-key report of Fridtjof Nansen, as retold by Marshall Gardner. This explorer made no exaggerated claims. In his diary, he dismissed the sighting he and his crew made of a dull-red sun during the polar winter as merely a mirage. And yet a solar light seen by many people at a time when it is supposed to be dark is not well accounted for by calling it a mirage.

The promise of a great rescue mission by space ships during a future Earth axis shift can be found in small booklets, with titles like "We Met the Space People," and also in full-page ads that appear in UFO-related magazines. I have speculated that the purpose of such a mission would be to collect quantities of blood and youth hormones like interferon, and possibly also organs for transplantation. If the planetary axis were to begin shifting, these UFO-related "prophecies" could be used to bolster the claim that the space people tried to warn us of the coming disaster, but most of us refused to listen. The truth, I believe, is that the so-called space people are the ones responsible for suppressing the credible warnings that our scientists would like to provide and that a few have provided on a small scale. The lack of any meaningful warnings delivered to the public ensures that when the axis shift does arrive, many of us will feel we have no choice but to accept the promised rescue by spaceships. Many, of course, will see this as a manifestation of the rapture prophesied in the Bible.

The idea that UFO forces may be intending to kill us is supported by a message that can be found in "Fobuserdeemus" (126), where one of Carl Sagan's characters mispronounces "Phobos or Diemos," the names of the moons of Mars. In Fobuserdeemus, we can find: "Some UFOs be redeemers, frends. Some be murderers."

The idea that space people might conceal a coming cataclysm from the people of Earth is portrayed in Isaac Asimov's novel *Nemesis*, as space people conceal the approach of a star toward the solar system. We have seen many riddles from *Nemesis* relating to the coming collapse of Mars and the threat that would be posed to Earth by the resulting fragments as well as the central star. (The riddle that follows is one of many that were discovered after I had almost wrapped up this book project.) In *The Norby Chronicles* Isaac and Janet Asimov portray the heroes traveling back in time to meet a couple of small dinosaurs. This would have been prior to the asteroid strike that wiped out the dinosaurs. The names of these dinosaurs provide a graphic message describing an approaching asteroid strike.

In "Ziphyzggtmtizm, Zargl" we can find: "Marz iz limp. Itz gapz (polar gaps) ar larg. Iph Marz phallz apart itz partz may graz Irth, may hit Irth." The fact that this riddle message comes from the names of two creatures of a type destroyed by an asteroid strike helps to confirm that the message is accurate. The letters for "Marz" and "larg" and "graz," as well as "itz," are grouped together. The words "iph" and "iz" and "ar" are given without the need to rearrange the letters.

This message goes into detail in describing the changes leading up to the collapse of Mars. The planet becomes limp without solar wind support for its rotation. The polar gaps get bigger as a result of less centrifugal force propelling matter toward the poles. The term "polar gap" was used in my early books, and so Asimov could assume that I might use the word gap in trying to solve this puzzle.

"Oola the All-Purpose Pet" is another creature encountered by Norby and his friends. Here we can find: "A polar hole shall shatter upper Europe." It is technically more accurate to say that the forces causing that hole to move will be responsible for shattering upper Europe. But riddles of this type do not lend themselves to long and complex messages.

"I peer at everi planet." This comes from Peter Valerian, an astronomer in Carl Sagan's novel *Contact*. As an astronomer, Sagan did indeed peer at the planets. He specialized in the study of Venus, but this message reminds us that he studied other planets as well. We have seen his prediction about Mars in "the Marshal Nedelin," which appears to be telling us: "Mars shall end and its inner star shall threaten the Earth." My safety as a hollow-Earth theorist was apparently a concern for Sagan. In "David Drumlin, Helga Bork" (126), he may have provided this message: "I am glad U are OK, R. A. Leedy. I believed U'd be murdered or have ur brain broken."

Isaac Asimov appears to have provided a similar message in W. Bradford Hume from "Where Is He?" (58). In this name, the message may be: "We feared U'd be harmed or murdered, R. A."

In Daneel Giskard Baley (66), Asimov predicts: "R. A. Leedy is a brain, kan learn riddle reading." Similarly, in "Manfred Dunkel, Euterpe Weiss" (51), we see: "Ewe kan learn to desifer mi pusles, R. A. Leedi." Also, in "Mabel, Bill, Gladys," from "In the Canyon" (30), Asimov predicts again: "Malleable as I am, Leedy may be able (to) see my messages." In riddles seen earlier, Asimov has described himself as mentally malleable, easily controlled by a superior power.

"Mars star wants to wander soon to a new area," Andrew Mortenson from "One Night of Song" (5).

"If the Mars star lands on Earth, the Earth shall die," Merrill "Handies" Foster (4).

"Mars appears to be a doomed planet. Cood Mars be saved?" This is from "Tapper Savand, Malcomber" (25).

"I am a real riddler, am making mani a riddle." This is from "Rikaine Delmarre, Gladia" (94).

"Do everi riddle over and over and ewe will learn." This is from "R. Daneel Olivaw" (96).

"Tri to steer the Mars star or it cood destroi the Earth." This is from "the Corridor-Master" (36).

The polar circle on this composite photograph of Venus was said by <u>Life</u> (Nov. 1980) to be "terra incognita, out of range of Pioneer's radar." In other words, the taxpayers paid for pictures of Venus and got pictures of only part of Venus, though other planets have been photographed from all angles. The construction of this model indicates a sloping down of land toward a polar basin or opening which has been covered over. For those who think satellite data proves that planets are solid at the poles, there is a picture of this same type to be found of Earth in <u>Scientific American</u> Jan. '89, p. 20. Beneath it is this comment: "The black circle represents a region outside the satellite's view."

There will be a mass landing of UFOs on the White House lawn. Leading television personalities will be revealed to be extraterrestrials. The poles will shift and UFOs will conduct a mass rescue of humanity from the disaster.

Does this sound familiar? These are the types of messages that mankind has been receiving from "space beings" in recent years. Perhaps this happens during a face-to-face UFO contact experience. More often it happens during a seance or other form of psychic contact with entities from "beyond". Dated predictions are given. Now and again they come true, just often enough to keep up our interest. But often we are disappointed. What is the meaning of false and confusing predictions? Is someone playing games with us?

In trying to answer these questions, our Ufologists have offered the following observations:

1. We cannot understand UFOs because their society is so far beyond ours. We are to them as ants are to us.

2. UFO society may be non-human in origin, leading to difficulties in communication.

3. UFOs may inhabit another dimension of reality, and so they would not communicate as we do.

4. Contactees may have misinterpreted the messages, due to cultural differences, language problems, etc.

With these excuses we make allowances for what often is a deliberate pattern of deceit. If message givers can speak clearly enough to tell you that there will be a mass landing of UFOs, and then this does not take place, can you conclude that you misunderstood, that their concept of time is different from ours? No, with a prediction like this which is clear and straightforward, when it does not come true, you have no choice but to believe the UFO information source has simply lied.

SOWING CONFUSION

Why would they lie? Rather than look to some complex sociological answer involving differences between them and us, we should look to a simple answer. Some might lie to us because they want us to disbelieve other UFO messages that are not lies.

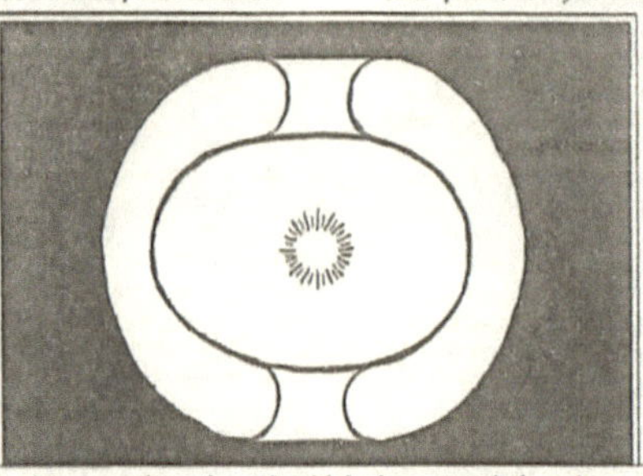

A cross section of a spheroidal planetary nebula, whose shape indicates what Earth must have looked like during its formation. This would also be the shape of a hollow planet.

WHEN UFO PROPHECIES DON'T COME TRUE

by
RUTH A. LEEDY

In other words, there is more than one faction in UFO society. Is that a surprising idea? Should we expect them all to think alike and have the same motivations? On the contrary! Some of them would care about us and give us useful messages, while others would regard us as a plague upon the planet and would try to confuse us so we would not benefit from any UFO messages.

The chances are that the faction or factions that care about us are not dominant. They may form a sort of underground movement, while the dominant regime is hostile to us. Why assume they are hostile? Because they do not make

(Continued next page)

official contact with our governments and because of hostile acts such as kidnapping contactees and mutilation of animals.

When a UFO message is received, how are we to know it is from a friend or enemy? No one can say for certain. However, you can assume that most UFO messages will be from the dominant UFO group.....our enemies. After all, they are more numerous and powerful than the UFO group friendly to us. They will feel more free to speak out.

USELESS ADVICE

A message that rambles on at great length spouting pious platitudes about how to conduct one's daily life— this message could be from the enemy. Why? Because it has a mesmerizing effect. It says nothing important and draws attention from other potentially more important UFO messages. If a great deal is said about how mankind is "spiritually unworthy" to receive certain knowledge or that humans need to have their consciousness raised about certain things, this again could be the enemy talking. Why so? Because this is mumbo jumbo. It is impossible for four billion people all to be simultaneously in a state of spiritual unworthiness. In any population, a certain percent will be spiritually advanced, while others are more backward. The "raising of consciousness" is a meaningless concept. One either has knowledge or one does not. When I go to the encyclopedia, I am not attempting to raise my consciousness about a subject. I am attempting to learn about it.

This so-called raising of consciousness can be seen as a substitute for knowledge. A UFO source may ramble on at great length about a coming polar shift with the stated intention of raising your consciousness about the subject. And yet the entire message may not contain a scrap of useful information about the coming disaster. A date may be given when the poles are supposed to shift, and that date may pass without incident. This leads most of the public to suspect that all UFO warnings on the subject are worthless. A few contactees will maintain that their consciousness has indeed been raised on the subject, but they will be unable to demonstrate that they have gained any solid knowledge about it.

If the UFO message gives information about unknown planets, this could be the enemy once again diverting our attention with useless trivia. Even if the planets Lanulos or Clarion really did exist, that would not have a great deal to do with life here on Earth. If the UFO spokesman claims to be from another galaxy, this is in all probability the enemy once again pulling our collective legs. The cost of space travel even to the nearest star, and the time involved, would make such visitation highly unlikely. Such statements draw our attention away from the fact that UFOs are a phenomenon of this solar system and this planet. They patrol here. They play here. They probably live close by if not in our midst.

WHAT WE NEED TO HEAR

What sorts of messages would our UFO friends be likely to give us? In other words, what is the information we really need as opposed to the useless trivia? Do we need information about various levels of consciousness, or how to conduct our lives, or how to learn about our past incarnations, or how to read the future? Do we need to know where the UFOs came from.....what unknown planet or galaxy? Some would say we need this knowledge. And yet, it is obvious that these same UFO guides who offer such pseudo-information also possess wondrous knowledge that they don't want us to have, secrets of space flight and life extension. What they offer us is tantamount to a missionary going to a tribe of starving, naked savages and offering only religious doctrine, without any attempt to bring the people food, medicine, and the trappings of civilization.

The only useful messages which UFOs are offering today, in my opinion, are messages about a polar shift which other UFO forces are attempting to keep secret. This is only my opinion; others will disagree. The number of UFO messages concerning a coming polar shift is truly staggering. Amidst this blizzard of signals, there must be a lot of meaningless noise. In other words, some of the warnings of disaster are ridiculous and nonsensical, probably to distract us from the useful warnings. If the message states that the sins of mankind are bringing on the coming disaster, or that the axis is becoming unstable due to "negative vibrations" from humanity, this is most definitely a useless message. The earth's axis does not shift for these reasons. Our atomic tests will not throw the axis off to any significant degree, though they do nothing to help it. The forces stabilizing the axis are strong enough to overcome the effects of an occasional atomic blast, or rocket launching.

THE TRUTH ABOUT DISASTER

What useful information should we look for in UFO warnings of a polar shift? Until one understands the polar shifting process, it is hard to answer that question. My studies have led to the conclusion that the disaster will involve an increase in earthquakes, tidal waves, a deluge, the rising and sinking of land masses, the shifting of all heavenly bodies, the darkening of the sun and moon, and the appearance of a mysterious light in the clouds. This may sound like and "end-time" prophecy, but actually I arrive at these conclusions by using the model of a hollow Earth and studying what would happen when the magnetic field reverses itself, as is predicted to happen within a few centuries. The formation of new magnetic poles would attract the powerful solar wind to strike at two new points, defining a new axis. The polar openings would be recarved or reformed to center at a new axis, due to centrifugal force. This would cause tidal waves and a spinning of ocean water into the atmosphere. This debris in the air would darken the sun and moon.

The rising and sinking of land masses would take place because of a thinning of the earth's shell at the location of the new equator. The mysterious light in

(Continued on page 43)

40

UFO PROPHECIES (Continued)

the cloud would be the Earth's interior sun shining through the new polar openings where it could be seen only by reflection in the clouds. The earthquake activity prior to the disaster would be the result of the radioactive heat increase within the Earth, the same heat increase that is causing our magnetic field to decline and eventually reverse itself. The shifting of heavenly bodies would take place because of a change in the planet's orientation in space.

If our scientists were doing their jobs, we would not need UFO warnings about this coming disaster. Our scientists would tell us all we need to know. And yet, through the centuries, these types of warnings have been passed on to us by "higher powers", as if they knew our scientists would not be free to tell us what was going to happen. To know of such a cover-up hundreds or thousands of years in advance would be to know of a UFO grand design by which the destiny of humankind was mapped out for centuries ahead of time.

PLENTY OF WARNING TIME

The coming magnetic field reversal has definitely been "in the works" for 2500 years, because that is how long the magnetic field has been declining. And that is about how long mankind has been receiving warnings of a coming global disaster. The prophecies of Christ about changes in the courses of heavenly bodies indicate he was predicting a polar shift *(Matthew 24:29)*. The follow-up prophecies of Edgar Cayce provide a stronger link between Christ's prophecies and a coming polar shift.

This is not an event that will take place on a particular day. It will be spread out over a period of weeks and probably months. The initial changes in the polar openings will darken the sky long before there is any major change in the axis.

So don't believe anyone who tells you the poles will shift on a particular day. It will take longer than that, and nobody knows the exact day when it will begin.

> *Ruth Leedy lives in Delaware and can be reached by writing to Route 3, Box 240-B, Dover, DE 19901. What a writer!!*

SEARCH MAGAZINE

Winter, 1991-92
Issue No. 189

Owl Press
P.O. Box 81
Rosholt, WI 54473-0081

JUDITH M. STATEZNY

March 31, 1992

More Reflections:

The message of Christ was essentially political. That is, the underlying message regarding the second coming was political. Therefore, naturally, it must be dealt with by the news media and political leaders.

I want no one to doubt my intention to be very hard on those who fail to cooperate. I put my life on the line here. In the past my approach may have been more diplomatic, more intellectual, and thus easier to ignore. Since it was easy to ignore, I was protected from reprisals. But I am shedding that diplomacy now and being totally direct. I insist this must succeed now. If it does not I may suffer consequences that I don't want to contemplate.

Yes, your names have been recorded. How you respond to this will be noted. If anything happens to me, that too will be noted, for there are those who see the matter as I do. When this material is mailed, more will see it. They will say, "My God, she took a risk here. Is she safe? Has anything happened to her?"

Now you are the ones who must take the risk. I am done. I've said it all over and over. If anything remains unclear, please contact me and more literature will be sent.

Best regards,

Ruth Leedy

Ruth A. Leedy

Hollow
Earth Mysteries
And the Polar Shift
by Floria Benton

Scientist finds tropical paradise inside the Earth:

There's a hole in the Pole

● FROM THE top of the world: Weather satellite picture showing view of the North Pole

UFOs come from inner space – not outer space!

"THERE'S A HOLE in the North Pole and it leads to a tropical paradise located at the center of the Earth."

That's the word from a highly respected scientist and explorer who led a research team to the top of the world last year. The mission was meant to check out astounding satellite photos revealing a large, dark opening at the North Pole.

"We have found a fabulous world," says Edmund Bork, who led a team of international explorers through the hole in the Pole last summer.

Tropical vegetation

"It has its own sun, a shallow, warm water sea and lush, tropical vegetation."

What's more, the land within is inhabited by a highly advanced and very peaceful race of humans. The Danish explorer refuses specific details on what he and the five other members of his team found. However, he says they were led to the 1,400-mile wide opening to the inner earth by a satellite photo released by the U.S. Department of Commerce in 1970.

The photo, taken in late November 1968 reveals a huge, black opening at the top of the world.

There is a similar hole at the South Pole, Bork claims.

"The Earth is shaped like a huge doughnut," he says. "We live on the outer side of the doughnut."

Normally, the hole cannot be seen from the air because of the heavy cloud cover over the North Pole, and because the inhabitants of the hollow Earth keep it covered with electronic "light screens."

The light screens give the illusion of vast fields of ice and snow through holographic manipulation of the real snow and ice that surrounds the opening, says Bork.

But the screens fail from time to time, especially during periods when the sun is in an intensive sunspot phase as it was in 1968.

"The hole is there and all we had to do was walk through the illusion," he says.

Because the hole is so large, the slope down is very gradual and the explorers were hardly aware they were entering another world.

Bork and his associates were elated, but not surprised, to find the Earth is hollow — and inhabited. Belief that the world is hollow is thousands of years old, he explains. "Legends of a hollow Earth and people who live there go back to the ancient Egyptians and Greeks. Even the American Indians believed people lived under our feet."

A number of other men claim to have entered the hole in the North Pole.

They include William Shavers, a Navy pilot who crashed at the North Pole during World War II, and a tribe of wandering Eskimos who told a Canadian reporter in 1958 that they had found "a green land at the top of the world." Leif Erickson and other Norse sea explorers claimed to have found a "lush green land" near the Arctic circle. "Later mapmakers thought they were talking about Greenland, but Greenland is hardly green, much less lush," Bork says.

Impressive evidence

Impressive evidence came with a radio transmission from Rear Admiral Richard E. Byrd, the great American polar explorer, as he flew over the North Pole in 1926.

"He said he saw below him a land of mountains, forests, green vegetation, lakes and rivers," says Bork.

"There was no mention of snow and ice, and the land he described belonged somewhere in the temperate zone, not at the North Pole.

"It sounded a lot like what the Norsemen described and what we discovered." — JOHN COFFIN

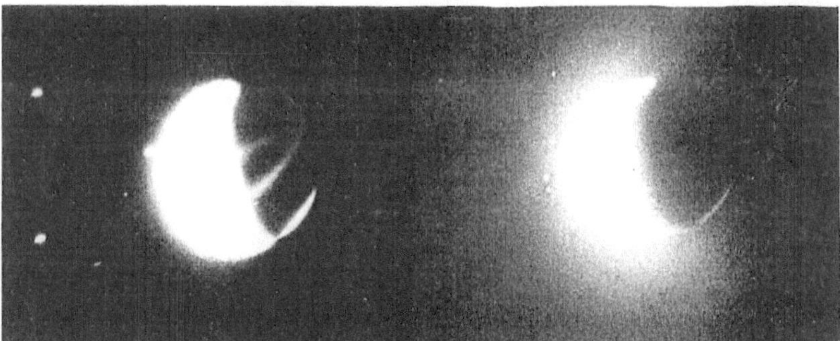

TWO VIEWS OF EARTH from Space Science and Astronomy, edited by Thornton and Lou Williams Page. The author of the captions evidently perceived the polar illumination which can be seen on the night side to be in the form of a solid cap of the type found on the Mariner 7 photographs of Mars. However, the aurora was revealed by polar photographs taken after publication of this book to be a ring of light.

FIG. 133. Far-UV photo of the Earth showing oxygen airglow, equatorial bands, and polar-aurora "caps," a 10-min. exposure taken from the Apollo-16 landing site on April 21, 1972, in light of wavelength 1230-1550A. The north pole is up and sunlight comes from the left. A blue star in Capricorn is very near the Earth's limb at left. (NRL-NASA photo.)

FIG. 134. Far-UV photo of the Earth showing the geocorona, a 15-sec exposure taken a few minutes before Figure 133 in light of wavelength 1050-1550A dominated by hydrogen Lyman-alpha (1216A). On longer exposures the geocorona extends over 40,000 mi. (NRL-NASA photo.)

"J. Jeppson helps hopeless people" Joseph Olsen (53). Asimov's wife, Janet Jeppson, was his psychiatrist prior to their marriage.

"A horned god harmed me" Gordon Hammer (53). Asimov's Azazel stories feature a demon who has horns and a tail.

"Everi riddle I have devised is an SOS sine." This is from Hari "Raven" Seldon (25).

An amazing series of three messages can be found in Asimov's story "The Fights of Spring" (5). These are as follows: "I cannot relax until U all see that there is a sun at the hollo Earth's center, at the axis. Cost will B high iph U all cannot C this. I hope mi book B (a) reel help." The first sentence comes from "Antiochus Schnell, Artaxerxes," the second from "Bullwhip Costigan," and the third from "Philomel Kribb."

"World in Danger" Riddle Echoes Novel

Earth is in peril in Asimov's novel *The Gods Themselves*. The danger is understood by no one and suspected by only a few when an unexpected type of warning is received. In the main character's name, we can find the words "be warned" rather easily. His name is Benjamin Allan Denison, and the name Andrew is added because another character calls him Benjamin Andrew Denison by mistake. In "Andrew" we can see "warned," and "be" is found at the beginning of the name. The complete message appears to be: "Be warned. World is in danjer. Small son inside Mars will soon wander."

Some warnings presented here have been clearer and stronger than others, and they have not come in the way we might have expected, but we have indeed been warned.

Afterword

Was Paul Broca a sacrificial figure, someone whose brain wound up in a jar, broken, because he tried too hard to sound cataclysmic warnings? Riddles cannot prove this, but they can suggest it.

"Branes are broken to keep secrets. Please share these messages on major catastrophes that are soon to come on hollo planet Earth." This message, which could also include "Pall Broca," can be found in "Herman Gelb, Peter Jonsbeck" (34). In this story, where Isaac Asimov portrayed potential nuclear destruction, the author *encoded three names* (Merry, Tessa, Ranay) from his cataclysmic novel *Nemesis* into the letters adjacent to commas and apostrophes.

"Pall Broka's broken brain is in a jar. My brain is nearly broken. I am in major pain." This message can be found in Carl Sagan's character names "Palmer Joss, Billy Jo Rankin" (126). One of these characters has a world map tattooed on his torso and was previously known as Geos the Earth Man, until (like Earth) he was struck by lightning.

"No savior hoo is in a jar is ani SOS," Vissarion Johnson (12). Here Asimov seems to indicate that heroic efforts by scientists to issue cataclysmic warnings to the public are no help if those potential saviors wind up in a jar.

"When Broca declared our world be hollow, he knew he could be murdered. When your broken brane be held by a jar, you no can redeem human race." This message can be found in "Henry Jarlow, Blanc and Nkrumah" (54), in a novel where Asimov portrays a star heading for Earth.

"Share messages on doomsday hidden in my riddles." This message can be found in "Sol R. Hadden, Yamagishi" (126), the names of two characters who are portrayed by Carl Sagan as orbiting the Earth together. These names can also say: "Inside hollo Mars a hidden son, or sol, is shining." Since "Sol" is given in the riddle, this statement is very likely meant to be added.

Two riddle messages seem to suggest that Asimov's riddles must be solved and published in a book in order to save the Earth. I really think that it can only be saved by many people's determination to save it, and that is what I hope to inspire in all of you.

Powerful earthquakes are said to cause the entire planet to vibrate or "ring" like a bell. The pattern of vibration in a hollow bell is probably quite different from the pattern of vibration in a solid structure. In his mystery "The Singing Bell," Isaac Asimov portrays a Dr. Wendell Urth as the owner of a hollow rocky bell, which is a natural formation and which he allows to be broken. The breaking of Dr. Urth's hollow rocky bell can be seen as a metaphor for a coming disturbance in the rocky structure of a hollow Earth.

Bibliography

Anderson, Flavia. *The Ancient Secret: In Search of the Holy Grail.* London: Victor Gollancz, 1953.

Angebert, Jean-Michel. *The Occult and the Third Reich.* New York: Macmillan, 1974.

Asimov, Isaac. *A Choice of Catastrophes.* New York: Simon & Schuster, 1979.

Asimov, Isaac. *A Whiff of Death.* New York: Doubleday, 1958.

Asimov, Isaac. *Azazel*; Doubleday, New York, 1988. Stories: (6) "A Matter of Principle;" (7) "Dashing Through the Snow;" (8) "Flight of Fancy;" (9) "Galatea;" (10) "He Travels the Fastest;" (11) "Logic Is Logic;" (12) "More Things in Heaven and Earth;" (13) "Saving Humanity;" (14) "The Dim Rumble;" (15) "The Evil Drink Does;" (16) "The Mind's Construction;" (17) "The Two-Centimeter Demon;" (18) "To the Victor;" (19) "Writing Time."

Asimov, Isaac. *Banquets of the Black Widowers.* New York: Doubleday, 1984. Stories: (21) "A Monday in April;" (22) "Can You Prove It?"

Asimov, Isaac. *Casebook of the Black Widowers.* New York: Doubleday, 1980. Story: "The Family Man."

Asimov, Isaac. *Fantastic Voyage II: Destination Brain.* New York: Doubleday, 1987.

Asimov, Isaac. *Forward the Foundation.* New York: Doubleday, 1993.

Asimov, Isaac. *Foundation.* New York: Doubleday, 1951.

Asimov, Isaac. *Foundation and Earth.* New York: Doubleday, 1986.

Asimov, Isaac. *Foundation and Empire.* New York: Doubleday, 1952.

Asimov, Isaac. *Foundation's Edge*. New York: Doubleday, 1982.

Asimov, Isaac. *Gold: The Final Science Fiction Collection*. New York: HarperCollins, 1995. Stories: (31) "Alexander the God;" (32) "Cal;" (33) "Fault-Intolerant;" (34) "Frustration;" (35) "Gold;" (36) "Hallucination;" (37) "Kid Brother;" (38) "Perfectly Formal" in "Cal;" (39) "The Nations in Space;" (40) "The Shiny Quarter" in "Cal."

Asimov, Isaac. *Magic: The Final Fantasy Collection*. New York: HarperCollins, 1996. Stories: (42) "Baby, It's Cold Outside;" (43) "It's a Job;" (44) "March Against the Foe;" (45) "Northwestward;" (46) "Prince Delightful and the Flameless Dragon;" (47) "The Critic on the Hearth;" (48) "The Fable of the Three Princes;" (49) "The Mad Scientist;" (50) "The Time Traveler;" (51) "To Your Health;" (52) "Wine Is a Mocker."

Asimov, Isaac. *Murder at the ABA*. New York: Doubleday, 1976.

Asimov, Isaac. *Nemesis*. New York: Doubleday, 1989.

Asimov, Isaac. *Past, Present and Future*. Buffalo: Prometheus, 1987.

Asimov, Isaac. *Pebble in the Sky*. New York: Doubleday, 1950.

Asimov, Isaac. *Prelude to Foundation*. New York: Doubleday, 1988.

Asimov, Isaac. *Puzzles of the Black Widowers*. New York: Doubleday, 1990. Stories: (59) "Sunset on the Water;" (60) "The Alibi;" (61) "The Fourth Homonym;" (62) "The Lucky Piece;" (63) "Unique Is Where You Find It."

Asimov, Isaac. *Robot Dreams*. New York: Doubleday, 1986. Story: "The Machine that Won the War."

Asimov, Isaac. *Robot Visions*. New York: Byron Preiss, 1990. Story: "Too Bad!"

Asimov, Isaac. *Robots and Empire*. New York: Doubleday, 1985.

Asimov, Isaac. *Second Foundation*. New York: Doubleday, 1953.

Asimov, Isaac. *Tales of the Black Widowers*. New York: Doubleday, 1974. Story: "The Obvious Factor."

Asimov, Isaac. *The Best Science Fiction of Isaac Asimov*. New York: Doubleday, 1986. Stories: (70) "A Loint of Paw;" (71) "Death of a Foy;" (72) "Eyes Do More Than See;" (73) "Franchise;" (74) "I Just Make Them Up, See!"; (75) "It's Such a Beautiful

Day;" (76) "Spell My Name With an S;" (77) "Strikebreaker;" (78) "Unto the Fourth Generation."

Asimov, Isaac. *The Caves of Steel*. New York: Doubleday, 1954.

Asimov, Isaac. *The Complete Stories, Vol. 1*. New York: HarperCollins, 1990. Stories: (81) "Gimmicks Three;" (82) "Old-fashioned;" (83) "Profession;" (84) "The Author's Ordeal;" (85) "The Dying Night."

Asimov, Isaac. *The Complete Stories, Vol. 2*. New York: HarperCollins, 1992. Stories: (87) "Lenny;" (88) "Light Verse;" (89) "The Key;" (90) "The Martian Way;" (91) "The Singing Bell."

Asimov, Isaac. *The End of Eternity*. New York: Doubleday, 1955.

Asimov, Isaac. *The Gods Themselves*. New York: Doubleday, 1972.

Asimov, Isaac. *The Naked Sun*. New York: Doubleday, 1957.

Asimov, Isaac. *The Norby Chronicles*. New York: Walker, 1984.

Asimov, Isaac. *The Robots of Dawn*. New York: Doubleday, 1983.

Asimov, Isaac. *The Union Club Mysteries*. New York: Doubleday, 1983. Story: "Gift."

Asimov, Isaac and Robert Silverberg. Nightfall. New York: Doubleday, 1990.

Asimov, Isaac and Robert Silverberg. *The Positronic Man*. New York: Doubleday, 1991.

Asimov, Isaac and Robert Silverberg. *The Ugly Little Boy*. New York: Doubleday, 1992.

Benton, Floria, pseudonym for Ruth Leedy Callinan and Ruth A. Leedy, now Ruth Leedy Carr. *Hollow Earth Apocalypse: Asimov's Warning*. Leedy, Wilmington, Del., 1985.

Benton, Floria, pseudonym for Ruth Leedy Callinan and Ruth A. Leedy, now Ruth Leedy Carr. *Hollow Earth Mysteries and the Polar Shift*. Saucerian, Clarksburg, W.Va., 1981.

Benton, Floria, pseudonym for Ruth Leedy Callinan and Ruth A. Leedy, now Ruth Leedy Carr. *The Hollow Earth at the End Time*. New Age, Jane Lew, W.Va., 1983 (also published under the author's name of Leedy).

Bernard, Raymond. *The Hollow Earth*. New York: Bell, 1969.

Blavatsky, Helen P. *The Secret Doctrine*. London: Theosophical Pub. Co., 1888.

Bloxam, Jeremy and David Gubbins. "The Evolution of the Earth's Magnetic Field." *Scientific American*, Vol. 261, No. 6, Dec. 1989.

Campbell, Joseph. *The Hero With A Thousand Faces*. Princeton: Princeton University Press, 1949.

Coffin, John. "There's a Hole in the Pole." *The Sun*. (tabloid), Feb. 25, 1992.

Emerson, Willis George. *The Smoky God*. Mundelein, Ill.: Palmer, 1965.

Encyclopedia Britannica. Chicago: Benton, 1974.

Gamow, George. *A Planet Called Earth*. New York: Viking, 1963.

Gardner, Marshall B. *A Journey to the Earth's Interior*. Aurora, Ill., 1920, reprinted by Health Research, Mokelumne Hill, Cal.

Gribbin, John and Stephen Plagemann. *The Jupiter Effect*. New York: Walker, 1974.

Heline, Corinne. *Mysteries of the Holy Grail*. Los Angeles: New Age, 1977.

Jackson, Joseph H. *Pictorial Guide to the Planets*. New York: Crowell, 1973. See title page.

Leedy, Ruth A. (See also Benton, Floria.) *Rebirth of a Planet*. Leedy, Wilmington, Del., 1983.

Life. "Lifting the Veil of Venus;" Vol. 3, No. 11, Nov. 1980.

Lowell, John and Kaspar Nureddin. *The Impending Golden Age*. Coolidge, Arizona: Upland Trails Press, 1958.

Marvin, Ursula B. *Continental Drift: Evolution of a Concept*. Washington: Smithsonian, 1973.

Moffat, Samuel and Elie A. Schneour. *Life Beyond the Earth*. New York: Scholastic Book Services, 1965.

Moore, Patrick. *Guide to Mars*. New York: Norton, 1978. (Key photograph omitted in a later edition.)

Mother Earth News. "Isaac Asimov: Science, Technology and Space." No. 65, Sept.–Oct. 1980.

Page, Thornton and Lou W. Page, Eds. *Space Science and Astronomy*. New York: Macmillan, 1974.

Pauwels, Louis and Jacques Bergier. *The Morning of the Magicians*. Stein & Day, 1977.

Sagan, Carl. *Broca's Brain*. New York: Random House, 1979.

Sagan, Carl. *Contact*. New York: Simon & Schuster, 1985.

Stearn, Jess. *Edgar Cayce: The Sleeping Prophet*. New York: Doubleday, 1967.

Sullivan, Walter. *Continents in Motion: The New Earth Debate*. New York: McGraw-Hill, 1974.

Trench, Brinsley LePoer. *Secret of the Ages: UFOs from Inside the Earth*. New York: Pinnacle, 1974.

About the Author

One of Ruth Leedy Carr's early ambitions was to be a science writer. While studying journalism at Indiana University, she specialized in science stories, in addition to taking a second major in psychology. She began researching the cataclysmic prophecies of Edgar Cayce in relation to the hollow Earth theory in 1980.

Her first two books were republished by Gray Barker of Saucerian Press, the original publisher of Raymond Bernard's classic, The Hollow Earth. She produced five books in the 1980s and has done numerous newsletters since that time. She was proclaimed by William L. Moore, a noted UFO researcher, to be the world's foremost proponent of the hollow Earth theory.